THE DIVINE INITIATIVE & THE HUMAN RESPONSE

NAVIGATING YOUR FAITH IN THE DIGITAL AGE

DWAINE AJ WHOGOES

DIRECTLIVING
PUBLISHING COMPANY
Sustainable Growth Through Creative

Book Cover by Cemebras
Illustrations by Dwaine AJ Whogoes
First edition 2024

TABLE OF CONTENTS

Hebrews 13:8

"Jesus Christ is the same yesterday and today and forever."

"In a world that rapidly changes with technological advancements, knowing that Jesus remains the same is reassuring. His teachings, love, and promises are enduring constants throughout our lives."

— PASTOR EMILY DAVID

THE DIVINE INITIATIVE AND THE HUMAN RESPONSE

The Divine loves humans so much that he gave his own life, opening the way for humans to become His children, for His desire is to share his life, His plan, and His creation was with His human family on earth.

Therefore, God Himself provided the human sacrifice on our behalf to atone for all human inadequacy, in which our mortality is plagued by physical death.

God's standard for life and righteousness is satisfied in Christ Jesus (The Divine); humans could never attain such a standard of life and righteousness on their own.

So then, faith in Christ Jesus' Death and Resurrection—the Divine Initiator—makes this divine quality of life and righteousness in God available to us, especially those who respond to His wondrous Gift of Grace through faith.

— DWAINE AJ WHOGOES

INTRODUCTION

How do we hold fast to our Christian faith in a world where a constant digital stream bombards us for attention? This question stands at the heart of our exploration as we seek to understand the delicate dance between divine sovereignty and our human responses in this rapidly evolving digital landscape.

This book's core is a journey—your journey and mine—through the challenges and opportunities the digital age presents to our faith. We are called to steer our spiritual path, understanding the initiatives God takes and how we are to respond. It's about finding balance and meaning in an era where screens often dictate more of our lives than our spirits.

I remember sitting in a small, dimly lit room several years ago, scrolling through my phone, feeling connected yet incredibly isolated. At that moment, I mentally crystallized my struggle with modern culture and technology—a constant tug-of-war between engaging with the digital world and my God. This personal battle sparked a deeper inquiry into how our faith can survive and thrive in such times.

Scripture is our compass in this exploration. Verses like Romans 8:28 remind us, "In all things, God works for the good of those who love him, who have been called according to his purpose." This scriptural foundation guides every page to deepen your understanding and equip you with the tools to live a faith-filled life amid digital distractions.

The structure of this book unfolds to aid your understanding and application of living a God-oriented life in a tech-saturated world. From personal reflections on divine sovereignty and human choice to practical steps for living out our faith in the age of smartphones and social media, each chapter builds on the last, aiming to foster a resilient, vibrant faith.

You, my readers, come from varied backgrounds and spiritual depths. Whether deeply entrenched in your faith or just beginning to explore what it means to live spiritually in a digital age, this book is for you. It invites all who seek to deepen their understanding of God's sovereignty and learn how to translate that into daily actions and decisions authentically.

As we embark on this journey together, I am filled with hope. Writing this book has been a pilgrimage of heart and mind, and I pray it will serve as a beacon for you. These pages offer encouragement, insight, and practical guidance to navigate your faith in this complex digital age.

May we all be equipped to face the challenges and embrace the opportunities that come with living faithfully in the digital era, inspired by the depth of Scripture, the authenticity of personal stories, and the practicality of guidance within these pages.

Matthew 6:33
"But seek first his kingdom and his righteousness, and all these things will be given to you as well."

"This verse powerfully reminds us to prioritize our spiritual lives over material and technological pursuits. While technology can add convenience and productivity to our daily lives, our primary focus should always be seeking God's kingdom.

When we put God first, everything else, including our technological endeavors, will fall into its rightful place. I often reflect on this when I get too absorbed in gadgets and screens."

— PASTOR EMILY DAVID

CHAPTER ONE

UNDERSTANDING DIVINE SOVEREIGNTY IN EVERYDAY LIFE

Have you ever wondered why specific challenges block your path, no matter which direction you turn? It's easy to view these obstacles as mere hindrances that disrupt our well-laid plans. However, what if there's more to these trials than meets the eye? What if they are divine interventions designed to shape, redirect, or prepare us for a future we've not yet imagined?

This chapter explores the profound relationship between divine sovereignty and our everyday experiences. You'll discover that your trials and tribulations aren't just random misfortunes but potentially pivotal moments where God's hand is at work. By understanding how to recognize these divine interventions, develop a gratitude mindset, and grow spiritually through difficulties, you'll find yourself better equipped to traverse the complexities of life with faith and purpose.

Isaiah 40:28

"Do you not know? Have you not heard? The Lord is the everlasting God, the Creator of the ends of the earth. He will not grow tired or weary, and his understanding no one can fathom."

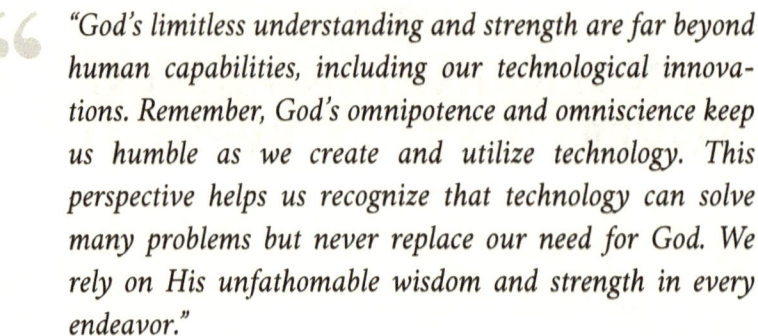

"God's limitless understanding and strength are far beyond human capabilities, including our technological innovations. Remember, God's omnipotence and omniscience keep us humble as we create and utilize technology. This perspective helps us recognize that technology can solve many problems but never replace our need for God. We rely on His unfathomable wisdom and strength in every endeavor."

— PASTOR EMILY DAVID

SEEING GOD'S HAND IN DAILY CHALLENGES

Psalm 19:1

"The heavens declare the glory of God; the skies proclaim the work of his hands."

Think about Joseph, a young man sold into slavery by his brothers, only to rise to become the second most powerful man in Egypt. His journey, filled with unjust treatment and years of hardship, might look like a series of unfortunate events from a purely human perspective. Yet, Joseph himself recognized these events as part of a larger plan, famously saying to his brothers, "You intended to harm me, but God intended it for good" (Genesis 50:20). His story is a powerful example of how God uses our difficulties for a greater purpose, often well beyond our immediate understanding.

By examining biblical narratives like Joseph's, we can learn to see our challenges through a different lens. Whether it's a job loss, illness, or relational strife, each challenge carries the potential to redirect your path or refine your character in ways that align with God's sovereign plan. Understanding this doesn't just change how you view your circumstances—it transforms how you live through them.

Developing a Gratitude Mindset

Adopting a mindset that looks for the good in all situations is not about denying the pain or difficulty of your circumstances.

Instead, it's about shifting your focus to what God might do through these challenges. This perspective is rooted in gratitude, fundamentally changing how we experience life's ups and downs.

Practicing gratitude, especially in tough times, can profoundly impact your spiritual health and overall well-being. Start by noting what you're thankful for daily, including the lessons learned from your struggles. This thankfulness helps cultivate a heart that readily sees God's hand at work.

Spiritual Growth Through Difficulties

James 1:2-4 tells us, "Consider it pure joy, my brothers and sisters, whenever you face trials of many kinds because you know that the testing of your faith produces perseverance." This Scripture isn't a call to enjoy suffering but an invitation to view our spiritual refinement as a joyous opportunity. Trials expose our weaknesses and areas needing growth, essential for maturing in our faith.

As you encounter various trials, consider them opportunities to deepen your trust in God and strengthen your spiritual resilience. Each difficulty is a chance to practice patience, humility, and faithfulness—qualities honed not in ease but in adversity.

Practical Steps to Discernment

Recognizing God's guidance in difficult situations requires intentional practice. Start by regularly setting aside time for prayer and reflection, asking God to reveal His hand in your circumstances. Keeping a spiritual journal can be incredibly helpful in this process. As you write down key events, thoughts, and feelings, look for patterns or repeated themes that may indicate where God is leading you.

Additionally, seek the counsel of spiritually mature friends or mentors who can offer biblical wisdom and objective advice. Their insights can help clarify God's direction when circumstances cloud your vision.

By embracing these practices, you set the stage for a life in which you not only steer challenges with grace but also grow wiser and more confident in God's sovereignty. As you continue reading, remember that each step taken in faith is a step toward a fuller understanding of how intricately and beautifully God is involved in the details of your life.

SOVEREIGNTY AND SERENDIPITY: FINDING DIVINE PURPOSE IN COINCIDENCES

In our lives, moments often arise that seem nothing short of serendipitous—those instances where you find just what you need or meet just who you need to meet, exactly when you need it most. To some, these might appear as happy coincidences; however, as we delve deeper into our faith, we begin to recognize these as 'God-incidences'—not mere products of chance but divine orchestrations crafted by our Creator for our benefit. This understanding transforms our perception of random events into moments of potential spiritual significance, encouraging us to consider their purpose in God's grand design.

Take, for instance, the story of Ruth in the Bible. Her decision to glean in Boaz's field may seem coincidental at first glance, but it marks a pivotal moment in biblical history. This "chance" encounter secured Ruth and her mother-in-law Naomi's survival and led her to become an ancestor of King David and, ultimately, Jesus Christ. It's a profound example of how God orchestrates seemingly random events into significant, life-altering encounters. Like Ruth, each "coincidental" meeting or opportunity could be a thread in a larger tapestry that God is weaving. Recognizing this can change how we approach every unexpected turn, viewing each as a moment of divine purpose.

We must cultivate a heightened spiritual sensitivity to truly appreciate these divine serendipities. This sensitivity can be fostered by maintaining a daily log of events that may initially seem coincidental. For example, jotting down unexpected encounters or conversations that proved meaningful later on can help us connect the dots retrospectively. Over time, this practice enhances our awareness of God's presence in our daily lives and reinforces our faith in His sovereign guidance.

The impact of recognizing and reflecting on these God incidences cannot be overstated. When we begin to see God's hand in what we once thought was random, our faith deepens, and our trust in His sovereign plan strengthens. This acknowledgment helps us to remain hopeful and patient during times of uncertainty, knowing that God has a purpose in every moment, whether or not we can see it immediately. Furthermore, sharing these personal stories of divine serendipity can encourage and uplift others in their faith, creating a ripple effect of spiritual growth and strengthened trust in God across our communities.

By embracing this perspective, we start to see our lives as an intricate puzzle that God is piecing together, where every piece has its place and purpose, no matter how small or puzzling. This shift in viewpoint enriches our spiritual journey and connects us more deeply with those around us as we collectively marvel at the mysterious and beautiful ways God moves in our lives.

BALANCING ACT: GOD'S CONTROL AND OUR EFFORTS

In the dance between divine sovereignty and human free will, we find ourselves in a realm of mystery where theology meets daily living. The Apostle Paul in Philippians 2:12-13 encourages us to "...work out your salvation with fear and trembling, for it is God who works in you, both to will and to work for his good pleasure." This passage beautifully encapsulates the dynamic interplay between God's overarching sovereignty and our responsibility. It suggests that while God is ultimately in control, orchestrating events beyond our comprehension, He also invites us to participate actively in His divine narrative.

Consider Moses, a Midian shepherd who encountered God in a burning bush. This moment was undeniably a sovereign intervention by God. Yet, it required Moses to make a series of decisions aligned with God's will, leading him back to Egypt to demand that the Pharaoh release God's people. Moses's journey was not just about accepting his destiny; it was about actively stepping into the role God had prepared for him, confronting his insecurities, and making choices that aligned with God's commands. His life illustrates how human actions and divine guidance coexist, each playing a pivotal role in the unfolding of God's plan.

Similarly, the story of Esther shows how human initiative works within divine sovereignty. Faced with the potential genocide of her people, Esther, a queen who could have easily remained passive, chose to act. Her decisions—to fast, pray, and strategically approach the king—were actions taken within God's sovereign framework, ultimately leading to the Jewish people's salvation. These narratives underscore that our efforts and decisions matter profoundly, even as they operate under God's will.

This understanding should spur us not into a passive resignation but into active participation in our faith and life decisions. God's sovereignty doesn't absolve us from action; instead, it empowers us to act boldly, knowing that our efforts are part of a larger divine tapestry. We are called to live responsibly and proactively, making decisions reflecting our faith while trusting God's overarching plan. This balance is pivotal, for leaning too far into passivity can lead us into fatalism—a belief that negates the significance of our actions and decisions, suggesting that behavior is irrelevant in the grand scheme.

Fatalism, the idea that all events are predetermined and, therefore, inevitable, is a misinterpretation of divine sovereignty. It strips us of our God-given choice and responsibility, leading to spiritual lethargy and disengagement. In contrast, Scripture teaches us that while God is sovereign, He has also endowed us with the ability to choose and act according to His will. Our actions are significant, and they interact with God's purposes in ways that are often beyond our understanding but are nonetheless real.

To cross this complex interplay, we need to cultivate a mindset of active faith. This starts with regular prayer and meditation on Scripture, seeking God's guidance for our lives and decisions. It involves surrounding ourselves with a community of faith that encourages and challenges us to live out our calling. It also means stepping out in faith, sometimes taking risks, as we make choices that align with our understanding of God's will.

Moreover, active participation in what God is doing requires discerning where He moves in our lives and joining Him in that work. It means seeing opportunities, seizing them, facing challenges, and embracing them as arenas for growth and testimony. In doing so, we honor God's sovereignty and our role within His divine initiative and plan, living not as passive observers but as active participants in the story He writes in and through our lives. This approach doesn't just change how we view our circumstances; it transforms how we live our lives, infusing our daily walk with purpose, passion, and a profound partnership alongside Him.

RESPONDING TO GOD'S INITIATIVES IN WORK AND FAMILY

In the intricate dance between our professional aspirations and personal lives, discerning God's calling can sometimes seem like looking for a clear signal in a storm. Yet, in these realms—our workplaces and homes—we find profound opportunities to live out our faith in tangible, impactful ways. Understanding God's initiatives in these aspects of our lives requires us to listen closely because faith comes by hearing God's word (Romans 10:17), observing the patterns in our experiences, and responding with intention and belief.

Recognizing God's call in our careers is not always about dramatic changes. Sometimes, it involves perceiving the divine purpose in our current roles or seeing a new opportunity that aligns more closely with our spiritual gifts and passions. For instance, consider a corporate lawyer who feels a pull toward working with a nonprofit that provides legal aid to the underprivileged. This shift may fulfill a professional desire and resonate deeply with their spiritual commitment to justice and mercy; principles heavily emphasized in Micah 6:8. To discern such calls, start by prayerfully considering your current job: What aspects bring you joy? Where do you find your skills and passions meeting the world's needs? Engage in conversations with trusted mentors who can provide guidance and perhaps identify blind spots or opportunities you may not have considered.

Balancing the demands of work and the call to worship God through our labor is another area requiring a divine calling. Integrating faith into the workplace goes beyond ethical conduct; it involves viewing our work as a form of worship, an offering to God. This perspective transforms our approach to daily tasks, turning routine projects into acts of service and excellence. For example, a teacher may view preparing lessons as part of her job and preparing to sow seeds of knowledge and wisdom in young minds, viewing each student as a precious life God entrusted to her care for that season. Such a mindset shifts our work from a mere job to a calling and mission, where the secular becomes sacred.

Family life, similarly, is a significant arena where God's work unfolds. It's where enduring relationships are built and where foundational values are imparted. Viewing family as a platform for God's work involves recognizing each member's role in God's plan. Parenting becomes an act of stewardship in which we help shape the next generation, not just in skills and knowledge but in character and faith. It's more than providing—it's modeling the fruits of the Spirit (Galatians 5:22-23), like love, patience, and kindness. When conflicts or challenges arise, they become opportunities to demonstrate forgiveness and grace. Each family activity, from shared meals to vacation times, can be infused intentionally, creating spaces where family members experience love, learn about God, and see their roles in His grand narrative.

The stories of individuals who have responded to God's call in their professional and family lives are both instructional and inspirational. Take the case of a small business owner who decided to implement ethical sourcing for his products after a conviction during a church service about fair trade and justice. This decision was challenging due to higher costs and, initially, lower profits. Still, it led to a loyal customer base that valued ethical practices, eventually growing the business more than before. Or consider a mother who, amid the busy life of raising three children, felt called to start a Bible study in her home. This small group grew into a vibrant community where many found faith and fellowship, significantly impacting her local church community.

These real-life examples underscore that responding to God's initiatives often involves practical, sometimes challenging choices that intertwine faith with everyday living. It consists of making decisions reflecting our commitment to following Christ, whether in boardrooms or living rooms. These choices set a powerful example for others, showing what living out one's faith authentically and wholeheartedly means.

As we seek to understand the demands and decisions of our careers and families, let us remain attentive to God's voice and leadership. Let us embrace the opportunities to serve, influence, and minister through our vocations and relationships. In doing so, we fulfill our calling to be salt and light in the world, influencing it for the better while fulfilling God's purposes. This active, intentional approach to living out our faith ensures that our work and family lives align with our spiritual beliefs and are vibrant expressions of them.

CHAPTER TWO

INTEGRATING FAITH WITH MODERN TECHNOLOGY

1 Corinthians 10:31

"So, whether you eat or drink or whatever you do, do it all for the glory of God."

How do we maintain a vibrant and authentic faith in an era where our lives are increasingly intertwined with digital platforms? This question echoes in many believers navigating the digital landscape. While technology offers unprecedented opportunities for connecting and learning, it also presents unique challenges that require us to rethink how we cultivate our spiritual lives. This chapter delves into the critical task of integrating faith with modern technology—ensuring that our digital engagements enrich rather than detract from our spiritual health.

GODLY NAVIGATION OF SOCIAL MEDIA

Setting Boundaries for Healthy Use

Navigating social media wisely demands intentional boundaries to safeguard our mental and spiritual well-being. It's akin to tending a garden; just as a gardener must fence off their plot to keep out pests that might destroy their plants, so must we set limits to protect our hearts (Proverbs 4:23). This begins with recognizing that not all content is beneficial. The apostle Paul reminds us in 1 Corinthians 10:23, "All things are lawful, but not all things are helpful." Applying this wisdom, consider setting specific times for social media use to prevent it from consuming your day or interrupting your moments of prayer and reflection. Tools like app timers can help you stick to these designated times.

Moreover, curating your feed ensures your consumption keeps your spiritual vitality healthy. Unfollow or mute accounts that stir discontent, envy, or anxiety, and instead, choose to follow those that inspire and uplift your spirit. It's also wise to regularly review your interactions and posts—ask yourself if they align with the fruits of the Spirit: love, joy, peace, forbearance, kindness, goodness, faithfulness, gentleness, and self-control (Galatians 5:22-23). If not, it may be time to make some changes.

Promoting Positive Engagement

When driven with purpose and positivity, social media can be a powerful platform for spreading the light of faith. It offers a unique space to share your testimony, biblical insights, or words of encouragement that can reach far beyond your immediate circle. Imagine the impact of a comforting Scripture posted during difficult times or a live video prayer session that others can join worldwide. These acts of digital kindness can be beacons of hope and sources of encouragement to others.

Moreover, engaging positively on social media involves active participation in conversations with grace and truth. It's more than just posting content—it's about interacting to reflect Christ's love. This could mean offering thoughtful responses to questions about faith, providing support in comment sections, or sharing resources to help others grow spiritually. Each interaction is an opportunity to be a digital disciple, cultivating a presence that honors God and edifies others.

Avoiding Digital Temptations

The anonymous nature of the internet can sometimes lead to temptations such as envy, pride, and gossip. These digital pitfalls are not new to the human condition but manifest uniquely online. James 3:5 compares the tongue to a tiny spark that can set a great forest on fire. In the digital age, this spark can be likened to a viral post or comment that can cause widespread harm. Vigilance is, therefore, essential.

When you feel tempted to engage in online debates that may lead to conflict or to scroll mindlessly, causing envy or dissatisfaction, take a moment to pause and reflect. Ask yourself whether this action will bring peace, strife, build up, or tear down. Implementing practical steps such as regular digital fasts can also help recalibrate your heart and mind. During these times, disconnect from digital devices and reconnect with God through prayer, meditation, and reading Scripture. This discipline helps strengthen your resilience against online temptations and keeps your focus on the things that truly matter to God.

Role Models of Faithful Social Media Use

Looking for role models who pilot social media gracefully and wisely can provide practical insights and inspiration. Consider the example of a well-known Christian leader who uses his platform to tackle tough questions about faith while maintaining a respectful and inviting demeanor. His posts are thoughtful, his interactions are kind, and his content always refers to Christ. Following such examples can inspire us to use our digital platforms for good.

Additionally, numerous Christian organizations leverage social media to serve communities, spread biblical teachings, and mobilize support for charitable causes. These groups often share stories of impact, post prayer requests, and provide updates on their missions, creating a sense of community and shared purpose. Observing and learning from these organizations can help you understand how to use social media as a tool for personal expression and as a platform for meaningful engagement and service.

Navigating social media with intention and integrity allows us to use these powerful tools to expand our reach, connect with others, and be a light in the digital world. As we continue to explore the integration of faith with modern technology, let us commit to using these platforms wisely, upholding our values, and positively influencing the digital space.

DIGITAL DISCIPLESHIP: OPPORTUNITIES AND PITFALLS

Proverbs 3:6
"In all your ways, submit to God; He will make your paths straight."

In this digital age, the landscape of discipleship is evolving rapidly, offering new avenues to deepen your faith in God through technology. Digital platforms have become a powerful tool for discipleship, providing access to various resources that can enrich your spiritual journey. Imagine the wealth of knowledge and spiritual guidance available at your fingertips—from Bible study apps that offer in-depth explanations and commentaries to online mentorship programs where seasoned believers guide newer believers in their faith walk. These tools are invaluable for believers who may not have ready access to physical church resources or are seeking flexible options to fit their hectic schedules.

For instance, consider a Bible study app that offers daily Scripture readings and includes video teachings and interactive forums where you can discuss the passages with others worldwide. This global network of believers can enrich your understanding and provide diverse perspectives to challenge and deepen your thinking about faith in God. Similarly, online mentorships, facilitated through platforms, connect you with more experienced Christians and provide personalized guidance and support, helping you understand personal challenges with biblical wisdom.

These digital tools empower you to take ownership of your spiritual growth, actively engaging with the content and applying it to your daily life.

So, while digital platforms are beneficial, you should maintain the irreplaceable value of personal relationships and community interactions in your spiritual growth. The early church thrived on face-to-face interactions, breaking bread, praying, and engaging in heartfelt discussions (Acts 2:42-47). A profound spiritual dynamic occurs when believers gather in physical fellowship—a dynamic that cannot be fully replicated online. Therefore, balancing your digital discipleship with regular involvement in local church activities is necessary. Participate in group studies, volunteer within community outreach programs, and engage in corporate worship. These activities foster a sense of belonging and accountability that digital platforms can complement but not replace.

Navigating the vast online content also requires discernment to ensure your teachings are biblically sound and doctrinally accurate. While a treasure trove of information, the internet also contains misleading and false teachings that can lead believers astray. To safeguard your faith, always vet the sources of your online spiritual content. Look for materials from credible organizations or individuals with a solid theological background. Check their doctrinal statements to ensure alignment with core Christian beliefs. Engaging with well-respected digital ministries known for their biblical fidelity can reassure you that the teachings you are consuming are grounded in Scripture.

Despite the benefits, there are pitfalls in digital discipleship that must be cautiously evaluated. The risk of isolating oneself in an echo chamber of similar thoughts and ideologies is high. Online platforms sometimes give us the illusion of being connected while allowing us to avoid the challenging aspects of a community, such as dealing with interpersonal conflicts or being accountable in a tangible way. To combat this, consciously engage in real-world church activities where you can be part of a living, breathing community. This engagement ensures that your faith is informed by digital input, tested, and lived out in daily interactions with other people.

Moreover, the convenience of digital tools can sometimes lead to passive consumption of spiritual content, where one might listen to sermons or participate in online prayers without actively integrating those teachings into personal life. To avoid this, set goals for your spiritual growth that require both online and offline activities. For example, if you use a prayer app, make it a point to journal your prayers by hand, reflecting deeply on your communication with God. Or, if you participate in an online Bible study, commit to discussing the lessons with a friend or mentor in person, which can enhance your understanding and application of the Scripture.

In essence, while digital discipleship opens up a new realm of possibilities for believers to grow in their faith, it should be monitored with wisdom and balance. By actively engaging with digital tools while profoundly investing in personal relationships and community life, you can ensure holistic spiritual growth that harnesses the best of both worlds.

USING TECHNOLOGY TO ENHANCE, NOT REPLACE, YOUR FAITH PRACTICES

Ecclesiastes 7:29
"This only have I found: God created mankind upright, but they have gone in search of many schemes."

Technology offers convenience and connectivity in the bustling rhythm of our daily lives. However, when it comes to our spiritual practices, it's vital to use technology to complement rather than substitute our engagement with God. Let's explore how we can integrate technology into our devotional life thoughtfully and effectively, ensuring it enhances our spiritual growth without becoming a crutch.

Integrating Apps into Daily Devotions

Consider the myriad of applications available designed to aid in spiritual disciplines. These tools can organize devotional readings, send prayer reminders, and help with scripture memorization. Still, it's essential to use them in ways that augment rather than diminish our personal devotion time. For instance, apps like YouVersion or Olive Tree provide daily Bible reading plans and devotional guides that help structure your spiritual routine. These apps can help you stay disciplined in your daily readings and reflections, especially if you have a hectic schedule.

However, while these apps offer convenience and structure, they should not replace the richness of meditating on God's word without digital interference. To strike a balance, you might read Scripture on an app during your commute and then spend some quiet time without digital devices to reflect on your reading. This practice allows you to benefit from the accessibility and organization of technology while ensuring that your engagement with Scripture is profound and meditative.

Another practical application is using prayer apps that remind you to pray at specific times or list prayer requests. These can be particularly helpful in maintaining a consistent prayer life. Yet, it's essential to remember that these prompts should serve as gentle nudges to turn your attention to God, not as the entirety of your prayer experience. Let these tools remind you to pause and connect with God, but ensure that most of your prayer time is spent in personal, undistracted communication with Him.

The Role of Podcasts and Videos in Spiritual Education

Multimedia resources such as podcasts and videos have become invaluable in our spiritual education, providing new dimensions of learning and engagement. These platforms offer a range of content —from sermons and biblical teachings to discussions on Christian living—that can complement your study of the Bible and enrich your understanding of its teachings.

For example, listening to a podcast from a reputable theologian can illuminate complex biblical passages or theological concepts, giving you a deeper understanding of your faith. Similarly, watching a video series on a book of the Bible can bring to light cultural, historical, and contextual elements that you might not grasp through reading alone. These resources are particularly beneficial in providing accessibility to those who learn better audibly or visually rather than through reading.

However, it's essential to approach these resources as supplements to your personal Bible study, not replacements. Use them to enhance your understanding and foster a richer engagement with Scripture, but ensure they are grounded in sound biblical doctrine and lead you to a deeper personal study of the Word. Always cross-reference what you learn with Scripture and consider discussing these insights with a mentor or study group to enhance your understanding and application further.

Virtual Church Services

The availability of online church services has been a significant development, especially highlighted when attending church physically was impossible. These services provide a way to stay connected with your church community and to participate in worshipping and teaching remotely. They can be a blessing for travelers, those dealing with illness, or those living in remote areas without a local church.

However, virtual services are best used temporarily or supplement physical church attendance. While they offer the sermon and perhaps an element of the worship experience, they lack the personal interaction and the communal aspects of a traditional church service—like fellowship, corporate worship, and communion—that are vital to a flourishing Christian life. If you regularly rely on virtual services, try to engage in some form of physical community, whether it's a home church group, small study gatherings, or serving in a local ministry. These real-life interactions are essential for building relationships and growing in faith in ways that virtual attendance can't fully replicate.

Technology as a Supplement, not a Substitute

As we commandeer the intersection of faith and technology, the overarching principle should be clear—use technology to enhance, not replace, your interaction with God and your involvement in a church community. Technology offers beautiful tools that can aid our spiritual practices, but it should never diminish our faith's personal and communal aspects. Let it be a servant, not a master, in your journey of faith, enhancing your spiritual growth and extending your reach in sharing the gospel, but always keeping the personal, heart-to-heart engagement with God and fellow believers as your primary focus.

VIRTUAL REALITY AND SPIRITUALITY: CAN THEY COEXIST?

Virtual reality (VR) technology, once a figment of science fiction, has become a tangible way to experience worlds beyond our physical reach. For believers, the potential of VR to enhance our understanding of biblical events or locations is particularly exciting. Imagine donning a VR headset and walking through the streets of ancient Jerusalem, witnessing the Sermon on the Mount, or exploring the inside of Noah's Ark. These immersive experiences could revolutionize how we connect with the Bible, making the stories more vivid and emotionally impactful. By stepping into a virtual recreation of biblical times, you're not just reading about history; you're experiencing it. This can lead to a more profound empathy for the figures within the Bible and a richer understanding of the context in which key events occurred.

However, as we integrate VR into our spiritual practices, we must handle this new territory with discernment. One of the primary ethical considerations is the potential for escapism. VR offers an escape to a different reality, but when used excessively, it may lead us away from engaging with our actual environment and community. The very essence of Christianity involves interaction with the world—serving others, building relationships, and being present in our communities. If we're not careful, the allure of virtual experiences can pull us away from these essential aspects of our faith.

Moreover, there's the risk of diluting personal interaction with God. Prayer and meditation have traditionally been intimate, personal practices that connect us directly with the Divine. If these practices are mediated through VR, we risk losing the personal touch that defines our relationship with God. It's important, therefore, to use VR as a tool that complements our spiritual practices without becoming a substitute for personal, unmediated worship and communion with God.

Balancing virtual and real-world spiritual practices is essential. While VR can supplement your spiritual education, it should not replace traditional, tangible practices such as attending church, participating in community service, or engaging in personal Bible study. Set clear boundaries for how and when to use VR to maintain this balance. It might involve allocating specific times for VR biblical tours or simulations, ensuring they are just one component of a diverse spiritual regime. Moreover, it encourages discussions about VR experiences in small group settings or church gatherings. This can help translate virtual experiences into deeper spiritual insights that can be shared and integrated into the community's life.

Looking toward the future, VR holds promising opportunities for spiritual education and communal worship. We may see more churches and spiritual organizations adopting VR to offer new forms of engagement, such as virtual reality church services for the homebound or detailed recreations of biblical events as educational tools. However, this also comes with challenges. The digital divide could become more pronounced, with those needing access to VR technology potentially feeling left out. Additionally, the novelty of VR may overshadow the deeper purpose of spiritual practices, turning them into entertainment rather than meaningful engagement with faith.

VR offers exciting opportunities and significant challenges as we continue exploring the convergence of technology and faith. Its potential to enrich our spiritual understanding is immense, but it must be approached carefully and focused on enhancing rather than replacing traditional faith practices. Let this technology serve as a bridge rather than a barrier, helping us connect more deeply with the spiritual truths of our faith while keeping us engaged in the tangible world around us.

In summary, as we delve into integrating VR into spiritual practices, it's clear that while the technology offers new avenues for experiencing biblical history and enhancing our understanding, it requires careful integration. Balancing its use with traditional spiritual disciplines ensures that our faith remains grounded and personal. As we move forward, it's exciting to consider how VR may continue to shape our spiritual landscapes, offering new ways to learn and connect while maintaining the core of our faith experience. The next chapter will explore technological advancements and their implications for personal and communal faith practices.

CHAPTER THREE

SPIRITUAL GROWTH IN A DIGITAL AGE

James 1:5

"If any of you lacks wisdom, you should ask God, who gives generously to all without finding fault, and it will be given to you."

How can we maintain a rooted, vibrant spiritual life in a world where digital connections often eclipse physical ones? This question isn't just theoretical; it's practical and personal. As we balance the intersections of faith and technology, adapting our holy life to the digital era is one aspect that calls for our attention. How can we keep our spiritual disciplines engaging and fruitful when so much of our lives are mediated through screens? This chapter invites you to explore the transformation of daily devotionals in this new digital landscape, ensuring that these timeless practices continue enriching your faith in the information age.

DAILY DEVOTIONALS IN THE DIGITAL ERA

Transitioning Traditional Devotionals to Digital Platforms

The transition of devotionals from traditional formats like printed books and live congregational teachings to digital platforms can feel like a leap into the unknown. Yet, this shift offers an opportunity to bring the age-old practice of daily devotion into the rhythm of modern life without losing its essence. Many believers now turn to apps and websites for their daily spiritual nourishment. Platforms like YouVersion or First5 offer well-structured, accessible devotional plans and insightful biblical commentary right at your fingertips. These digital tools are designed to fit seamlessly into your daily routine, whether reading a morning devotion over coffee or listening to a reflective passage on your commute.

The key to successful digital devotionals lies in their ability to preserve the core elements of traditional devotion—Scripture reading, reflection, and prayer—while using the benefits of technology. For instance, these platforms often include features allowing you to bookmark favorite passages, create notes, or customize reading plans according to your spiritual needs. This transition maintains the essence of devotion and enhances it, making engagement with Scripture more interactive and personalized.

Customizing Devotional Content

One of the most significant advantages of digital devotionals is their customization ability. Unlike a one-size-fits-all printed devotional, digital platforms can offer a variety of reading plans that cater to different spiritual seasons, theological interests, or personal preferences. Whether you're looking to dive deep into the Psalms, study the life of Paul, or explore themes like grace or forgiveness, these platforms can tailor content to meet your needs.

This customization extends to learning styles as well. Some people may benefit from reading reflective passages, while others could engage more deeply through audio or visual content. Digital devotionals can meet these varied needs with multimedia options, enhancing understanding and retention of Scripture. For example, the Dwell app offers an auditory experience of the Bible, allowing you to listen to Scriptures in different voices and translations, which can be a profound way to let the Word of Christ dwell in you richly (Colossians 3:16).

Consistency and Accessibility

Digital devotionals shine in their ability to help us maintain consistency in our spiritual growth. Maintaining regular spiritual disciplines can be challenging in our fast-paced, often chaotic life schedules. Digital platforms mitigate this by making devotionals easily accessible anywhere and anytime. Whether on a smartphone, tablet, or computer, these resources are just a click away, ensuring you can connect to God's Word regardless of location.

Moreover, the flexibility of timing offered by digital platforms means that you can engage with your devotional at a time that suits you best, which is convenient for developing a habit that sticks. Whether you are an early riser or a night owl, digital devotionals are ready when you are, making it easier to build and sustain a rhythm of daily engagement with Scripture.

Interactive Features to Enhance Engagement

Many digital platforms incorporate interactive elements stimulating engagement and reflection to deepen the holy experience. Features like reflection questions or prayer prompts invite you to read and respond to the Scripture you've studied. For instance, after a morning devotional, you might encounter a question that prompts you to reflect on applying a biblical principle in your day, or you may be encouraged to write down a prayer in response to what you've read.

Moreover, community features such as public comments or group discussions can enrich your understanding of Scripture. Engaging with the insights and experiences of fellow believers can broaden your perspective and encourage you in your faith journey. Some apps also allow you to share prayer requests or praise reports, fostering community and mutual support among users worldwide.

These interactive features not only make daily devotionals more engaging but also more impactful. They encourage you to read God's Word and meditate on it, discuss it, and live it out, fulfilling the Psalmist's exhortation to delight in the law of the Lord and meditate on it day and night (Psalm 1:2).

As we process our spiritual growth in this digital age, embracing these digital tools for daily devotionals offers a powerful way to keep our faith vibrant and active. While technology advances, our goal remains constant—to deepen our relationship with God through daily engagement with His Word. Let these digital platforms be your companion in this endeavor, enhancing your devotional life with accessibility, customization, and interactive engagement.

PRAYER APPS AND THEIR EFFECTIVENESS

In the landscape of digital spirituality, prayer apps present a fascinating fusion of ancient practice and modern technology, offering tools that can significantly enhance our prayer lives. Understanding the features that contribute to the usefulness of these apps can help you choose and use them more wisely, ensuring they enhance rather than burden your spiritual journey.

Features of Effective Prayer Apps

Successful prayer apps often share core features designed to support a disciplined approach to prayer. Reminder settings are the most straightforward yet powerful tools these apps offer. By setting up notifications, you can receive gentle nudges throughout your day, prompting you to pause and connect with God. This can be especially helpful in maintaining a rhythm of prayer amid the busyness of daily life. Another invaluable feature is prayer tracking, which allows you to log your prayer requests and note when and how they are answered. This can be a profound encouragement as you visibly track God's faithfulness. Moreover, options for communal prayer lists enable you to share prayer requests with friends or church groups, fostering a sense of community and mutual support in your prayer life. This shared approach enhances your commitment to prayer and deepens your communal bonds as you and your peers intercede for one another and celebrate answered prayers together.

Comparative Analysis of Popular Prayer Apps

Navigating the plethora of available prayer apps can be daunting. Apps like Echo, PrayerMate, and Abide have risen in popularity, each offering unique features that cater to different aspects of prayer life. Echo, for instance, is highly regarded for its user-friendly interface and robust reminder system, making it an excellent tool for those needing structure in their prayer habits. However, some users find its social sharing options limited, which could be a drawback for those who value communal engagement. PrayerMate offers extensive customizability, allowing users to create and organize prayer lists easily; its ability to import prayer points from various sources is a standout feature. Yet, its interface may feel less intuitive to those less tech-savvy. Abide stands out for its guided prayers and meditations, which are excellent for users seeking a more contemplative prayer experience. Its subscription-based model could be a barrier for users seeking a free resource.

Integrating Prayer Apps into Daily Life

Integrating prayer apps into your daily routine is vital to benefit from them genuinely. Start by setting realistic goals for your prayer life and use the app's features to help you meet them. Set reminders for morning, midday, and evening if you aim to pray thrice daily. Customize these times to fit your schedule, aligning them with daily activities like morning commutes or lunch breaks, making it easier to establish a consistent habit. Also, the tracking features should be actively used to maintain a list of ongoing prayer requests. Update these as situations evolve, and regularly review answered prayers to nurture a heart of gratitude. For communal prayer features, consider setting up a weekly review of shared requests within your prayer group, perhaps through a group chat or during meetups, to reinforce community ties and shared commitment to prayer.

Assessing the Impact on Personal Prayer Life

The real test of any prayer app's value lies in its impact on your personal prayer life. Studies and personal testimonies suggest that users of prayer apps often experience an increase in their prayer frequency and a more profound sense of connection in their spiritual lives. For instance, a user might share how the regular reminders helped transform their sporadic prayer habits into a robust daily routine, profoundly deepening their relationship with God. However, while many users find these apps beneficial, there is a cautionary note to consider—dependency. It's vital to ensure that the use of an app remains a facilitation rather than a crutch. If you feel obligated to the app rather than drawn to prayer, it may be time to re-evaluate your usage. Balancing digital aids with the natural flow of spiritual disciplines ensures that your prayer life remains personal and heartfelt, rooted in genuine communion with God rather than in a mere digital routine.

By exploring these facets of prayer apps, you can harness their potential to enrich your prayer life and times, making your daily conversations with God more disciplined, deep, and communal. As you watch your spiritual practices in this digital age, let these tools draw you nearer to the heart of God, enhancing your prayer life with the eternal yet ever-present words of faith spoken in quiet corners of your heart and this digitally connected world.

PODCASTS AS A TOOL FOR SPIRITUAL ENLIGHTENMENT

In the expansive universe of digital media, podcasts stand out as both a beacon of knowledge and a companion for your spiritual journey. Choosing suitable podcasts is foundational as you seek to deepen your understanding of God's Word and its application to your life. Not all spiritual podcasts are created equal; the key is to find those that offer sound theological content and align with your spiritual growth goals. To explore this choice, identify podcast hosts respected within the Christian community for their integrity and depth of biblical knowledge. Reputable pastors, theologians, and seasoned Christian leaders often host podcasts that provide rich, scripturally grounded content. Look for podcasts that cite their sources and base their discussions on biblical Scripture, ensuring that what you're listening to will enhance, not confuse your faith.

When choosing a podcast, consider what you hope to achieve spiritually. Are you seeking daily encouragement, theological depth, or practical advice on living out your faith? For instance, if you're interested in understanding biblical texts more deeply, seek out podcasts that focus on biblical exegesis or those hosted by biblical scholars. These can provide insights into the Scriptures' historical context, literary structures, and theological themes. Podcasts like *The Bible Project* offer comprehensive, accessible explorations of biblical books and themes that can enrich your personal Bible study. They break down complex ideas into manageable, engaging discussions that clarify rather than complicate your understanding of the Bible.

Moreover, podcasts can be a powerful tool for community learning, transforming a typically solitary activity into a communal experience. Listening to a podcast episode on a particular Scripture or topic with your small group or family can spark rich discussions and diverse interpretations that enhance everyone's understanding. Set aside time in your group meetings or family devotions to listen together and discuss the episode. This shared experience not only fosters a deeper communal bond but also allows you to benefit from the insights and reflections of others, which can broaden your perspective and push you to think more deeply about your beliefs.

However, while podcasts offer a wealth of knowledge and convenience, they should complement rather than replace other spiritual practices. There's a risk of becoming too reliant on passive listening as your primary spiritual input. To counter this, engage actively with what you hear. Take notes as you listen, jotting insights, questions, and reflections. This not only helps to reinforce what you learn but also enables you to integrate these insights into your personal Bible study. After listening to an episode, open your Bible and explore the passages discussed. See if the podcast's insights change or deepen your understanding of the text. You could also follow up your listening with prayer, asking God to help you apply the podcast's teachings to your life. This active engagement ensures that the podcast is a tool that genuinely enriches your spiritual growth rather than merely another form of entertainment.

Navigating the rich landscape of spiritual podcasts can profoundly enrich your journey with Christ. You harness what this modern tool offers by choosing wisely, integrating listening into your community activities, and balancing it with active engagement. Let each episode guide you a step further in your understanding of God's Word, and let the shared experiences of listening draw you closer to fellow believers as you grow together in wisdom and faith.

THE ROLE OF ONLINE COMMUNITIES IN SPIRITUAL SUPPORT

The digital age has ushered in unprecedented ways to connect, and among them is the burgeoning growth of online spiritual communities. These virtual gatherings provide unique advantages, particularly their ability to transcend geographical boundaries, bringing together individuals from diverse backgrounds to share, learn, and grow in faith. Online communities can be a spiritual lifeline for those with mobility issues or those living in remote areas, offering access to fellowship and teaching where traditional church attendance might be challenging or impossible. Furthermore, the internet's vast reach allows for the formation of niche groups that cater to specific theological interests or life situations, such as support groups for Christians battling illness or forums for those interested in more profound theological studies. This customization ensures that everyone can find a community that resonates deeply with their spiritual needs and personal circumstances.

However, engaging in these digital communities has its challenges. The lack of physical presence can sometimes lead to a sense of isolation or superficial connections, which lack the depth and warmth of face-to-face interactions. Anonymity, while beneficial in some respects, can also pose risks, as it may encourage dishonesty or reduce accountability among members. Moreover, the vast and unregulated nature of the internet necessitates a high degree of discernment. Only some online spiritual groups hold to sound doctrinal principles, and the impersonal nature of online interactions can sometimes allow for the spread of misinformation or unhealthy theological views.

Adopting best participation practices is a choice for navigating these waters safely and effectively. It begins with choosing the right communities by looking for groups aligned with sound biblical doctrine and fostering healthy, respectful, and informative conversations. Once you join, remember that the etiquette you'd use in physical interactions applies here, too—be kind, respectful, and truthful. Protecting your personal information is also paramount; be cautious about sharing sensitive details until you are confident in the community's credibility and the integrity of its members. Contributing positively involves more than just participating in discussions; it's about building others up through prayer, encouragement, and sharing insights to benefit the whole group.

Several case studies highlight the profound impact these communities can have. Consider the story of Sarah, a young mother in a rural area who found spiritual growth and support through an online group for Christian parents. The group provided friendship, encouragement, valuable resources, and advice that helped her walk through the complex challenges of parenthood with faith and grace. Then there's John, who struggled with addiction and found an online support group that was instrumental in his recovery. The group's availability and anonymity provided him with a safe space to seek help, leading to meaningful connections with others who guided and prayed for him through his journey to sobriety.

These stories underscore the transformative potential of online spiritual communities. They supplement our faith journey and can be vital platforms for encouragement, learning, and connection. As we embrace digital tools and platforms, let these communities serve as beacons of light, offering support and fellowship to those who may otherwise feel alone in their spiritual walk.

As we wrap up this exploration of online communities, remember the broader theme of this chapter—the role of digital tools in fostering spiritual growth. Online communities, prayer apps, podcasts, and digital devotionals are all part of a more extensive toolkit available to us in the digital age. Each tool has unique strengths and pitfalls, but collectively, they can significantly enrich our spiritual understanding. As we transition to the next chapter, we'll delve deeper into how these tools interact with traditional religious practices. We will explore ways to integrate digital and traditional forms of worship and community. This ongoing synthesis is a dance of striving to live out our faith authentically and practically in a rapidly changing world.

CHAPTER FOUR

THE HUMAN RESPONSE TO DIVINE ACTION

H ave you ever stood at a crossroads, feeling the weight of each potential path pressing upon your heart and mind? In these moments, amid the quiet tension of choice, we find ourselves wrestling with the profound interplay of divine sovereignty and human free will. How do we discern God's will when options sprawl before us like branches on a tree, each leading to a different outcome and future? This chapter delves into the dynamic dance of making choices that align with God's will, illuminating the role of divine guidance and personal freedom in shaping the tapestry of our lives.

FREE WILL AND FAITH: MAKING CHOICES ALIGNED WITH
GOD'S WILL

Daniel 12:4
"But you, Daniel, roll up and seal the words of the scroll until the end.
Many will go here and there to increase in knowledge."

Understanding the Interplay Between Divine Guidance and Free Will

The delicate balance between God's sovereignty and our free will is a cornerstone of theological reflection and a pivotal aspect of daily faith. Scripture offers numerous examples where divine guidance and human decision-making converge, painting a picture of a sovereign God intimately involved in human affairs. Consider the narrative of the apostles in the Book of Acts, where decisions about missionary journeys and church disputes were made under the guidance of the Holy Spirit yet involved active human planning and decision-making. These accounts highlight that while God orchestrates the grand narrative, He invites us to participate actively, making choices that resonate with His will.

This partnership does not mean that God controls every decision we make. Instead, He allows us to choose within the framework of His overarching purposes. Understanding this dynamic requires us to recognize that our choices matter deeply, both in their immediate impact and ripple effects on our spiritual journey and beyond. We must choose life with the awareness that our freedom is a gift meant to be exercised under the guidance of His Spirit, aligning our desires and decisions with God's larger plan for our lives and others.

Practical Tips for Discerning God's Will

Discerning God's will is not about uncovering a secret plan hidden behind a divine curtain but rather about aligning our hearts and minds with His. We begin this alignment through prayer, a fundamental practice that turns our spirit towards God's frequency. When faced with decisions, make prayer your first response, not your last resort. Ask God to clear your paths and fill your heart with peace regarding the best choice.

Seeking counsel from spiritually mature individuals is another vital step. Proverbs 11:14 says, "In an abundance of counselors, there is safety." Whether it's a pastor, a trusted friend, or a family member whose faith you admire, these individuals can offer biblical wisdom and objective advice to clarify God's direction in your situation.

Scripture is a lamp unto our feet and a light unto our path (Psalm 119:105). When making decisions, immerse yourself in God's Word. The principles and precepts found in the Bible provide a solid foundation for making wise choices. Confirm your decisions against Scripture; if a choice aligns with the truths of the Bible and promotes peace, love, and justice, it is likely in line with God's will.

The Role of the Holy Spirit in Decision-Making

John 14:16-17
"And I will ask the Father, and he will give you another advocate to help you and be with you forever — the Spirit of truth. The world cannot accept him, because it neither sees him nor knows him. But you know him, for he lives with you and will be in you."

The Holy Spirit is not a distant, abstract force but a present and active personality in the life of every believer. In John 16:13, Jesus describes the Holy Spirit as the Spirit of truth, who guides us into all truth. This guidance is necessary for decision-making, as the Holy Spirit works within us to shape our desires, convince our hearts, and prompt our actions according to God's will.

Be attentive to the Holy Spirit's promptings. These promptings may be persistent thoughts, unease about a particular decision, or a sense of peace about another. Cultivate sensitivity to the Holy Spirit through regular prayer, meditation on Scripture, and openness to His guidance. Our relationship to the Spirit is fundamental to living a life responsive to God's divine initiatives and aligning our will with His sovereign plan.

Case Studies on Moral Choices

Real-life scenarios often provide the most explicit illustrations of abstract concepts. Consider the case of Anna, a marketing executive who was faced with the choice to promote a product she knew was misleading. Her decision involved not just a career move but a moral dilemma. Through prayer, consultation with mature Christian friends, and reflection on biblical principles of honesty and integrity, she voiced her concerns to her team, advocating for transparency. This decision not only aligned with her Christian values but also led to a long-term positive impact on her company's reputation.

Then there's the story of David, a college student, deciding whether to join a mission trip or take a prestigious internship. Through seeking the Holy Spirit's guidance, engaging in Scripture, and discussing with his mentors, David felt a clear call to participate in the mission. The experience deepened his faith and opened doors to new opportunities to serve in ways aligned with his spiritual gifts.

These examples underline the significance of our choices and the profound impact they can have on our lives and the lives of others. As you face your own decisions, big or small, remember that each choice is an opportunity to live out your faith in practical ways, responding to God's divine action with human initiative that is thoughtful, intentional, and aligned with His will.

ACTIONS SPEAK LOUDER: LIVING OUT FAITH IN PUBLIC

Proverbs 18:15

"The heart of the discerning acquires knowledge, for the ears of the wise seek it out."

Boldness in Public Faith Expressions

In a world often marked by a reluctance to discuss faith openly, embracing the courage to express your beliefs publicly is not only bold but necessary. As Christians, we're called to "shine like stars in the sky" as we "hold firmly to the word of life" (Philippians 2:15-16). This metaphor isn't just poetic; it's a directive to live out our faith visibly and vibrantly. It's about letting others see the transformative power of Christ in our lives through our actions, words, and choices. Think of it as wearing your faith on your sleeve—showing others that your actions are rooted in a deeper, spiritual soil.

Christians do not brandish what we believe to provoke or confront but rather live in such a way that our lives speak volumes of the grace and love of Christ. This means being a beacon of kindness in a harsh office environment, offering forgiveness when it's easier to hold a grudge, or standing up for justice when it might be more comfortable to remain silent. Each action draws a picture for the world, vividly depicting what it means to live according to Christ's teachings.

Engaging in public faith involves verbal witnessing when opportunities arise. It's being ready to answer anyone who asks the reason for your hope but doing so with gentleness and respect, as 1 Peter 3:15 advises. Whether it's sharing how your faith has helped you overcome personal struggles or how it shapes your daily decisions, these personal anecdotes are powerful in touching the hearts and minds of people around you.

Balancing Professionalism and Faith

Navigating the intersection of faith and professionalism in the workplace can sometimes feel like a tightrope. On one side, you desire to live out your faith openly; on the other, you must maintain professional decorum and respect amongst workplace norms. Striking this balance requires wisdom, discernment, and a clear understanding of your environment and the legalities that govern who you are.

For starters, it's essential to recognize that expressing your faith at work doesn't mean proselytizing at every coffee break or turning every meeting into a Bible study. Instead, it's about embodying Christian principles through your work ethic and interactions with integrity. It's being fair in your dealings, honest in your communications, and compassionate toward your colleagues. These qualities often speak louder than words and open doors to deeper conversations about your faith in Christ in appropriate contexts.

However, challenges may arise, especially in secular or diverse workplaces where expressions of faith are viewed with skepticism or even hostility. Navigating sensitively and respecting others' beliefs is fundamental in such environments. This might mean choosing the right moment to share your faith or expressing it more through your actions than words. Always ensure that your expressions of faith adhere to workplace policies and cultural norms, avoiding any actions perceived as coercive or discriminatory.

Community Service as Faith in Action

Community service is a powerful avenue for living out your faith in tangible ways. It's putting your faith into action and serving others with words and deeds. Jesus Himself said that He came not to be served but to serve (Mark 10:45), setting the ultimate example for us. Engaging in community service allows you to mirror this aspect of Christ's character, showing love and compassion in practical ways that meet the needs of others.

This can be as simple as volunteering at a local food bank, participating in community cleanup days, or offering your skills to a nonprofit organization. Each act of service, no matter how small, reflects God's love and can profoundly impact those around you. Furthermore, these activities can serve as a bridge, connecting your church with the broader community and opening up opportunities for dialogue about your faith.

To integrate serving into your routine, identify causes you are passionate about and find organizations that align with those interests. Many churches partner with local charities and can provide volunteers with opportunities to serve. Alternatively, you could initiate a service project, rallying friends, family, and church members to address a specific need in your community.

Examples From Public Figures

Looking to public figures who have successfully integrated their faith into their professional and public lives can provide inspiration and practical insights. Consider the example of Eric Liddell, famously known as the "Flying Scotsman." He was an Olympic gold medalist and a devout Christian who refused to run on Sunday, standing firm in his convictions despite immense pressure and the international spotlight. His story, popularized in the film Chariots of Fire, illustrates the power of living out one's faith regardless of the impact of public scrutiny.

Another contemporary example is Denzel Washington, an acclaimed actor who speaks openly about his Christian faith and its influence on his life and career. He uses his platform to inspire others, often discussing the importance of faith, prayer, and humility in his interviews and public appearances. His willingness to integrate his faith with his public persona offers a model for how public figures can use their influence to bear witness to their faith in respectful and impactful ways.

These stories show the feasibility of living out faith in the public eye and the profound influence such a lifestyle can have. They encourage us, regardless of our sphere of influence, to live authentically and boldly for Christ, using our platforms, no matter how big or small, to reflect God's grace and truth in the world.

RESPONDING TO CONVICTION: CHANGE OF HEART, CHANGE OF ACTION

Romans 12:2
"Do not conform to the pattern of this world but be transformed by renewing your mind. Then you will be able to test and approve what God's will is—his good, pleasing, and perfect will."

Have you ever felt a deep, undeniable stir within your spirit, a sense that God is calling you to make a significant change in your life? This sensation, often called spiritual conviction, is a profound experience in which the Holy Spirit communicates directly to your heart in an urge to align more closely with the thing God wants.

Understanding this process is vital for believers seeking to grow and mature in their faith. Repentance is a spiritual conviction that can range from an urging to abandon a particular sin to responding to the call toward a new direction in life, such as a career change or a new ministry engagement. This isn't merely about feeling guilty or remorseful; it's about experiencing the realization that drives us to lasting change and transformation.

When we are convicted of sin, it's not to condemn or shame us, but God lovingly redirects our path to one that leads to greater joy, fulfillment, and alignment with His divine life purposes for our lives. For instance, consider the moment Peter, after denying Jesus three times, locked eyes with the Lord. The profound guilt he felt was not the end of the story—it was the beginning of a transformative journey that led him to become a pillar of the early church. His conviction led to repentance and, ultimately, to a bold, new chapter of his life dedicated to spreading the Gospel of Christ.

Turning spiritual conviction into actionable steps is essential to genuine transformation. The first step is usually confession, acknowledging the truths of our conviction before God. This confession is both a release from sin and a commitment to walk with our Creator. Following this, seeking forgiveness and cleansing is vital, as stated in 1 John 1:9, which promises that if we confess our sins, God is faithful and will forgive us and cleanse us from all unrighteousness. These initial steps pave the way for practical changes in behavior, thought patterns, or life choices.

For example, if you repent from an unhealthy habit and then replace it with a wholesome one, this is a practical step toward transformation. Suppose the issue is spending too much time on social media, leading to neglect of spiritual duties or personal relationships. In response, you might limit your social media use to certain times of the day, filling the reclaimed time with prayer, Bible study, or meaningful interaction with loved ones. This change doesn't just adjust a schedule; it realigns your daily life more closely with your values and spiritual goals.

Role of Accountability in Maintaining Changes

Maintaining the changes inspired by spiritual conviction often requires more than initial enthusiasm—a structured approach to accountability. Accountability involves inviting one or more trusted individuals into your transformation journey to provide support, encouragement, and, sometimes, gentle correction. These partners serve as external checks and reminders of your commitment to change, helping sustain your momentum long after the initial conviction may feel less intense.

Essential accountability structures or small groups can be set up in various ways, depending on the nature of spiritual maturity and growth the group agrees to work toward. For instance, if you're working on implementing significant periods of Bible study into your daily routine, an accountability partner could be someone who commits to weekly checking-ins to discuss what you've learned. Alternatively, you could join or form a small group focusing on common areas of Christian principles, such as purity, sobriety, or financial stewardship. The small group can provide a platform for sharing struggles and victories, offering accountability, mutual engagement, and encouragement.

Moreover, leveraging technology enhances accountability. Various apps allow sharing your goals and progress with accountability partners or groups. Whether a daily check-in via a mobile app or a shared online journal documenting your journey, these tools can help keep you honest and on track.

Testimonies of Changed Lives

Hearing stories from those who have walked this path can be incredibly inspiring. Take, for example, the story of Eve, a young professional who felt prompted by the Holy Scriptures about her passive involvement in her local church. This inspiration led her to volunteer as a youth leader, a role that helped her grow in her faith and profoundly impacted the teens she mentored. Eve's decision to act on her conviction transformed her spiritual life and had ripple effects in her community, illustrating the expansive impact of an open heart to God's promptings.

Then there's John, who was prompted by the Holy Spirit about his lack of generosity and changed how he approached his finances. He decided to start tithing regularly and set up a fund to support local charities and his local church assembly, which altered his financial management and deepened his reliance on and trust in God's provision. His testimony is a powerful reminder of how God uses our obedience to bring about personal growth and to bless others.

These stories underscore the transformative power of responding to what God wants to say and what He wants to do for us with concrete actions. They remind us that while the process can be challenging, the outcomes are invariably worth it, leading to deeper faith, more robust character, and a more impactful Christian witness. As you reflect on your experiences of repentance and change, consider how you might inspire others with your story, sharing how God has worked to bring about spiritual renewal and personal growth in you and to those around you.

WHEN GOD CALLS: STORIES OF RADICAL OBEDIENCE

1 Timothy 6:20-21

"Timothy, guard what has been entrusted to your care. Turn away from godless chatter and the opposing ideas of what is falsely called knowledge, which some have professed and in so doing have departed from the faith."

Defining Radical Obedience

In our faith walk, there are times when God calls us to go beyond the routine of daily obedience—to step out in radical obedience. This kind of obedience goes beyond the typical responses to God's commands; it involves a profound surrender, often requiring significant sacrifice and deep trust in God's plans for our lives. Radical obedience is marked by doing what is asked of us and embracing God's will wholeheartedly, even when it leads us into the unknown or demands a price we never thought we'd be called to pay.

The essence of radical obedience lies in its depth and the extent of the sacrifice involved. It's about following God's call without reservation, often in ways that defy human logic or personal comfort. This obedience is a vivid demonstration of our faith in His Word, a testimony to our belief in God's sovereignty and goodness, no matter the circumstances. It's the kind of commitment that says, "Not my will, but Yours be done," echoing Jesus's submission to the Father in the garden of Gethsemane as he sweats blood falling to the ground (Luke 22:44).

Biblical Examples of Radical Obedience

The Bible is replete with stories of individuals who exhibited radical obedience. Take Abraham, for instance, who was called to leave his country, his people, and his father's household to go to an unknown land that God would show him. This act of obedience, leaving everything familiar behind, was radical. Yet, there's perhaps no more a profound example of his obedience to God than when he was asked to sacrifice his only son, Isaac. The story in Genesis 22 is not just about the willingness to surrender what was most precious to him but about trusting God's goodness and provision even when it seemed like all hope was lost.

Then there's Jonah, whose story contrasts initial resistance with eventual radical obedience. Jonah's mission to Nineveh was daunting; he was called to preach repentance to a city notorious for its wickedness—so overwhelming a task that he initially ran the other way. However, after a dramatic series of events, including being swallowed by a giant fish, Jonah obeyed God radically. He went to Nineveh and delivered God's message, which led to the city's repentance and salvation from destruction.

These stories are ancient narratives and potent reminders of the transformative impact of radically obeying God. These examples challenge us to consider how deep our obedience goes and what it might feel like to trust God completely, even when His commands seem daunting or the outcomes uncertain.

Contemporary Examples of Radical Obedience

In our modern context, radical obedience is still very much alive. Consider the story of a young couple who felt called to adopt multiple children from overseas despite the enormous financial and logistical challenges. Their journey was fraught with obstacles, yet their commitment to follow God's call led to the formation of a beautifully diverse family, each child a living testimony to God's provision and love.

Another example is a businessman who felt compelled to sell his lucrative company to fund a recovery program for people with addiction issues in his community. This decision involved financial sacrifice and the risk of stepping into an entirely new, challenging area of ministry. However, the impact of his obedience has been profound, with hundreds of lives transformed through the program he helped establish.

These examples underscore radical obedience, not grand gestures in isolation but a sustained commitment to following God's lead, no matter the cost. They show us that when God calls, He also provides—whether it's the strength to persevere, the resources needed to carry out His commands, or the peace that comes from knowing you are following His will.

Encouraging Readers to Step Out in Faith

As you reflect on these stories of radical obedience, consider what God might call you to do. Is there an area in your life where you feel a deep, perhaps even uncomfortable, push toward a higher level of commitment or a new direction? Stepping out in faith could mean changing your career path to align more closely with your God-given passions, or it may involve reaching out to a neighbor or co-worker with whom you've wanted to share the gospel but who you've hesitated to approach.

Embracing this call starts with prayer, knocking on doors, and courage from God. Joshua 1:7-8 says: "Be strong and very courageous. Be careful to obey all the laws my servant Moses gave you: do not turn from it to the right or the left, and you may be successful wherever you go. Keep this Book of the Law always on your lips; meditate on it day and night so that you may be careful to do everything written in it. Then, you will be prosperous. This process may also involve community—sharing the concerns of your heart with wise counsel and seeking support from fellow believers who can pray and encourage you. Remember, radical obedience is not a solitary venture; it's a journey supported by the strength of our faith in God and the community we belong.

As you consider stepping out, remember that radical obedience is ultimately about God's love. It's a response to the profound love He has shown us—a deep love that led Him to send His only son to die for us. Our obedience reflects our trust in His love, a testament to our faith in His good and perfect will.

Let these stories and principles resonate in your heart as you close this chapter. Reflect on how the bold steps of obedience in your own life could transform you and have a ripple effect in your surroundings, bearing witness to the power and love of God. As we move forward, let us do so with an open heart to God's divine call, ready to act in radical obedience and eager to see how our lives unfold in the light of His holy purpose.

CHAPTER FIVE

BUILDING GODLY CHARACTER IN A SECULAR WORLD

Matthew 28:19-20

"Therefore, go and make disciples of all nations, baptizing them in the name of the Father, Son, and the Holy Spirit, and teaching them to obey everything I have commanded you. And surely, I am with you, always to the very end of the age."

How do we cultivate a life marked by godliness and virtue in a world of instant gratification and self-promotion? It's like being a gardener in an arid desert, constantly nurturing and watering the seeds of virtue amid the harsh winds of societal vices. This chapter is dedicated to understanding this daily battle between virtues and vices, offering you practical guidance on cultivating a character that not only withstands but also thrives amid the challenges of modern society.

VIRTUES VERSUS VICES: A DAILY BATTLE

Genesis 11:6

"The Lord said, 'If as one people speaking the same language, they have begun to do this, then nothing they plan to do will be impossible for them.'"

"This verse, spoken in the context of the Tower of Babel, shows the potential of unified human endeavor. Modern technology is a testament to what we can achieve collectively. However, these efforts can lead to pride and self-sufficiency without God's guidance. It's a reminder that we should pursue technological advancements while remaining humble and reliant on God."

— PASTOR EMILY DAVID

Today's culture often glamorizes or trivializes human offenses or sins that are deeply destructive to personal character and societal well-being. Greed, envy, pride, and slothfulness are not just personal failings; they are societal epidemics that subtly influence who we are and decision-making on a massive scale. Greed, for instance, drives relentless accumulation at the expense of ethical considerations, seen in corporate scandals or personal lifestyles that prioritize accumulation over generosity. Envy corrupts contentment, leading to a culture of constant comparison and dissatisfaction, amplified by the highlight reels of social media. Pride fosters division rather than unity, pushing individuals to focus on self-promotion rather than community building. Sloth, often overlooked, manifests not just in physical laziness but in a reluctance to engage deeply with issues, leading to superficial understandings and half-hearted commitments.

These vices often operate under the radar, woven into the fabric of everyday interactions and decisions. Recognizing them requires vigilance and a deep understanding of their manifestations and consequences. This awareness is the first step in countering their influence in your life and fostering a culture that upholds and honors virtues.

Cultivating Virtues as Counteractions

Cultivating virtues may seem daunting in the face of these prevailing vices, but it is both possible and essential. Scripture provides profound guidance on this front. For instance, generosity directly counters greed, encouraging a spirit of giving and sharing that aligns with the teachings of Jesus, who said, "It is more blessed to give than to receive" (Acts 20:35). Cultivating generosity can start with simple, intentional acts of sharing your time, resources, or talents with those in need.

Contentment stands against envy, a virtue that allows us to live in peace with our circumstances and fosters gratitude for what we have. Paul's words in Philippians 4:11-13, where he speaks of learning to be content in any situation, provide a robust framework for developing this virtue. Practicing contentment involves a deliberate focus on gratitude and a commitment to valuing relationships and spiritual riches over material gain.

Humility, which counters pride, involves recognizing our limitations and the value of others. It is grounded in the recognition of God's greatness and our dependence on Him, as well as the acknowledgment of the worth and contributions of others. Daily practices of humility include listening more than speaking, seeking advice, and celebrating others' successes as if they were your own.

Lastly, diligence opposes sloth; it means cultivating a consistent and earnest effort to fulfill our duties. Proverbs 13:4 highlights, "The soul of the sluggard craves and gets nothing, while the soul of the diligent is richly supplied." This can be practiced through setting personal goals, maintaining a disciplined schedule, and committing to continual learning and improvement, whether in personal, professional, or spiritual realms.

Scriptural Foundations for Virtue

The development of these virtues is deeply rooted in Scripture, which offers commands and the living examples of Jesus and the apostles. For instance, Jesus' life was marked by profound humility, service, and sacrifice—characteristics that starkly contrast with the religious leaders of His time, who often displayed pride and greed. The apostles, too, exemplified virtues like diligence in their missionary work and contentment even in persecution.

Reflecting on these scriptural foundations provides both inspiration and practical guidance. It connects the virtues we strive to cultivate with the larger narrative of God's kingdom, where these qualities are personal improvements and testimonies of a life transformed by the Gospel.

Daily Practices to Enhance Virtue Cultivation

Cultivating virtue requires daily commitment and practice. It's about making consistent choices that align with Christ's character and implementing practical steps that integrate these virtues into everyday life. Prayer is fundamental in this process, serving as a daily touchstone for divine strength and guidance. Incorporating Scripture meditation into your daily routine can reinforce your commitment to virtue, as the Word of God has the power to transform your mind and actions.

Though often overlooked, fasting is an assertive discipline for cultivating self-control and humility. It helps you detach from worldly dependencies and focus more on spiritual matters. Community service is a practical outlet for virtues like generosity and diligence, providing opportunities to live out your faith in tangible ways that benefit others.

By embedding these practices into your daily life, you create a rhythm of virtue that not only counters the vices prevalent in society but also shapes you into a beacon of God's light and love in the world. As we walk the intricacies of modern living, let these practices guide us, transforming our character and influencing those around us in profound and lasting ways. (Matthew 11:28)

PATIENCE AND PERSISTENCE IN A FAST-PACED WORLD

The Virtue of Patience in Modern Life

In today's world, where the pace of life accelerates daily, and the demand for instant results seems unending, cultivating the virtue of patience can feel like swimming against a powerful current. Our society prioritizes speed and efficiency, valuing quick decisions and rapid turnover. This cultural climate can make patience seem outdated, like a relic of a slower, less connected era. Yet, as Scripture highlights, the benefits of patience—peace, wisdom, and improved decision-making—remain profoundly relevant. Patience allows us to endure difficult circumstances gracefully, make thoughtful decisions, and maintain our integrity when facing challenges. It involves waiting actively, with a sense of purpose and trust, and understanding that some of the best things in life come through a process of slow and steady growth.

Scripture is replete with admonitions to embrace patience as a critical component of a faithful life. Consider James 1:4, which urges believers to let patience have its perfect work, that they may be perfect and complete, lacking nothing. This passage suggests that patience is not passive waiting but an active engagement in spiritual maturation. In a world that often confuses haste with success, patience allows us to see life's challenges at a pace that preserves our well-being and respects the natural rhythm of growth and change. By embracing patience, you resist cultural pressure and open yourself up to peace and wisdom from moving through life deliberately and thoughtfully.

Building Persistence Amid Instant Gratification

Closely linked to patience is the virtue of persistence—the ability to keep moving forward, steadily and firmly, toward a goal, especially when faced with obstacles or delays. In an age of instant gratification, where the temporary and the transient are often more celebrated than the enduring, persistence is a quality that requires both courage and resilience. It means maintaining your course even when the immediate rewards are not visible, trusting in the value of what lies at the end of your perseverance.

Developing persistence is foundational in areas of spiritual growth and community engagement. These aspects of life inherently require time and sustained effort—they cannot be rushed or instantly downloaded. For instance, deepening your spiritual understanding or building genuine community connections are endeavors that unfold over months and years. They are built through small, daily acts of faithfulness—choosing to spend time in prayer and Scripture study, committing to regular fellowship, and consistently serving others. These are not tasks to be checked off a list but practices to be woven into the fabric of everyday life, enriching your spiritual journey and strengthening your connections with those around you.

Practical Exercises for Developing Patience and Persistence

Consider integrating specific exercises and practices into your routine to cultivate these virtues. Mindfulness techniques, for example, can be particularly successful in fostering patience. By training yourself to be fully present in the moment without rushing to the next item on your agenda, you cultivate an awareness that helps temper the impulse for immediate results. Start with a simple daily habit of spending five minutes in quiet reflection, focusing on your breathing and the sensations around you, letting go of the past and future, and grounding yourself in the now.

Deliberate exposure to situations that require patience can also be instructively challenging. For example, taking on a long-term project, like gardening or learning a new skill, demands patience and persistence. Such activities teach you to value the process over instant results and help build resilience against frustration and discouragement.

Reflective journaling is another powerful tool. Regularly writing down your thoughts on your progress, setbacks, and feelings can provide insights into your growth in patience and persistence. This practice encourages a reflective approach to life's challenges and victories, helping you see the value in each step of your journey, regardless of the pace.

Stories of Patience Leading to Success

Real-life stories and biblical narratives underscore the transformative power of patience and persistence. Consider the story of Joseph in the Bible, who endured years of hardship, from slavery to imprisonment, before his situation turned around dramatically, leading him to become a ruler in Egypt. His story is a powerful testament to the fact that patience and persistence, rooted in faith, can lead to profound personal and spiritual rewards.

Consider someone like John, a community leader committed to revitalizing a troubled neighborhood in contemporary times. His efforts required immense patience and persistence as he faced initial resistance and slow progress. However, his steadfast dedication significantly improved the community's infrastructure and social cohesion. His story is a testament to the impact one person's persistent effort, guided by patience and a vision for betterment, can have on a community.

These ancient and modern stories remind us that patience and persistence are ethical niceties and practical necessities that can lead to substantial success and fulfillment. They challenge us to look beyond the immediate and the useful to the longer and often more rewarding paths that require us to wait, endure, and keep moving forward with purpose and faith.

HUMILITY IN ACHIEVEMENTS: GIVING GOD THE GLORY

1 Peter 4:10

"Each of you should use whatever gift you have received to serve others as faithful stewards of God's grace in its various forms."

Humility is often misunderstood today and is frequently associated with weakness or lack of confidence. However, from a biblical perspective, humility recognizes that our gifts, talent, and success are not merely our own doing but are granted by God's grace. It knows our strengths and our limitations, acknowledging that while we plant and water the seeds, it is God who makes them grow (1 Corinthians 3:7). This view shifts our perspective from self-centered pride to God-centered gratitude, allowing us to handle both our achievements and our failures with grace.

Maintaining humility can be incredibly challenging for personal and professional success. The world often urges us to claim credit, build our brand, and use our accomplishments as a platform for self-promotion. However, the scriptural call to humility teaches us to view our successes differently. For instance, when Daniel interpreted King Nebuchadnezzar's dream (Daniel 2:27-30), he didn't claim the wisdom as his own but explicitly credited God for the revelation. Following such a model means consistently acknowledging the source of our talents and successes. One practical way to do this is by publicly praising God and those who have contributed to our achievements. This may involve mentioning colleagues in a presentation, sharing success stories highlighting team efforts rather than individual prowess, or giving testimonies in your community or church about how God has enabled your accomplishments.

Continual self-assessment in light of Scripture is another vital strategy for maintaining humility. Regularly reflecting on passages such as Philippians 2:3 ("Do nothing out of selfish ambition or vain conceit. Rather, in humility value others above yourselves") can keep our hearts aligned with God's expectations. This practice helps cultivate a spirit of humility by reminding us of Jesus's example of servanthood. Serving others through mentoring, volunteering, or simply lending a hand in daily tasks can also help keep our egos in check, reminding us that we are part of a community and that leadership is, first and foremost, a service role.

The Role of Community in Fostering Humility

Community plays a basic role in fostering and maintaining humility. Being part of a faith community where humility is valued can provide a model to emulate and a gentle corrective when pride starts to creep in. In such communities, members can hold each other accountable, not through judgment or criticism but through honest, loving dialogue that encourages growth. For example, a small group or an accountability partner can help you reflect on your actions and attitudes, pointing out areas where pride may overshadow humility, and in repentance, we pray for a heart more like Christ's.

This communal aspect of humility is beautifully illustrated in how the early church operated, sharing everything they had and supporting one another (Acts 2:44-47). Such radical sharing and communal living are extreme examples, but they underscore that our lives and gifts are not solely our own; everything we have ultimately belongs to God. Therefore, they are meant to be used for the common good. By actively participating in community life—listening to others, sharing our resources, accepting help when needed—we learn to live out humility in practical, everyday ways.

Examples of Humble Leaders

The Bible offers numerous examples of leaders who demonstrated humility in their roles. Moses, described as "very humble, more than anyone else on the face of the earth" (Numbers 12:3), led Israel not through forceful dominance but through faithful service and dependence on God. In the New Testament, Paul, despite his vast theological knowledge and pivotal role in the church, consistently credited his success to God and described himself as the least of the apostles (1 Corinthians 15:9).

In contemporary settings, we see similar traits in leaders prioritizing service over status. Consider a CEO who foregoes bonuses to ensure fair wages for all employees or a pastor who spends as much time counseling parishioners as preaching. These leaders embody humility in their words and, more importantly, in their actions. They recognize their role as stewards rather than masters, consistently pointing others to God rather than seeking their glory.

As you grow, look to these examples as models for humility in success. Consider how your accomplishments can be used for personal gain, serving others, and glorifying God. Remember, humility isn't about diminishing your worth but recognizing the source of your strength and success. In doing so, you keep your soul in check and lead by example, showing others the power of a spirit-filled life with genuine humility.

CHARITY IN THE AGE OF SELF-PROMOTION

In today's society, where the spotlight often shines on achievement and personal branding, charity can seem like a whisper in a storm. The cultural push toward self-promotion and individual gain challenges the very essence of charitable living, which calls for selflessness and a focus on the well-being of others. It's a tug-of-war between uplifting oneself and lifting others. This cultural backdrop makes believers need to understand and embrace the biblical call to love—not just as an occasional duty but as a heartfelt lifestyle.

As depicted in the Bible, love encompasses far more than financial donations. It's about a holistic approach to serving people, including your time, abilities, and capacity for forgiveness and advocacy. The parable of the Good Samaritan (Luke 10:25-37) stretches the concept of charity beyond monetary help; it involves noticing the needs around us, crossing societal boundaries to help others, and taking action that costs us something, whether time, resources, or comfort. This story challenges us to look beyond our circles and address the needs of those who are often overlooked or avoided.

Cultivating a charitable spirit in today's self-centered world begins with intentional actions. Setting regular commitments to volunteer, for example, can anchor your week with a reminder of the needs around you. Whether it's helping at a local food bank, tutoring children after school, or visiting older adults, these commitments help integrate the practice of giving into your routine, making charity a natural part of your life rather than an afterthought. Furthermore, engaging in community aid programs addresses immediate needs and builds connections within the community, fostering a network of support and mutual care.

Promoting causes that benefit the wider community rather than oneself is another practical way to live out charity. This may involve using social media platforms to highlight and support charitable organizations, advocating for community projects, or organizing fundraising events. These actions help direct resources where needed and shift the focus from self-promotion to community upliftment.

The impact of living a charitable life is profound, both for the recipients and for you, the giver. On a personal level, engaging in charity can infuse your life with a more profound sense of purpose and fulfillment. It's about connecting to something larger than yourself and finding joy and satisfaction in seeing others thrive. Psychologically, generosity and service contribute to personal well-being, often enhancing the giver's happiness and life satisfaction.

The effects of charity on the community are equally transformative. Regular charitable activities can strengthen community bonds, reduce isolation, and create a culture of reciprocity and care. In communities where members look out for each other and support one another, resilience and solidarity can grow, making the community better equipped to face challenges together. Moreover, charity can lead to social change, addressing systemic issues that contribute to suffering and injustice, thereby improving the overall health and well-being of the community.

As we wrap up this exploration of charity in the context of a self-promoting era, remember that each act of giving, no matter how small, contributes to a larger tapestry of generosity that can transform both individual lives and entire communities. Let the practice of charity be a beacon that guides you through the fog of self-centered pursuits, leading you into a life marked by generosity, purpose, and deep fulfillment. As we turn the page, let's carry this spirit of giving into every area of our lives, allowing it to shape how we interact with the world and perceive and respond to our needs.

TRANSFORM TECHNOLOGY FROM A TEMPTATION TO A VITAL SPIRITUAL TOOL

"If God did not want us to discover something--raw materials or natural laws or potential powers--he simply didn't code it into the pattern of his creation."

— TONY REINKE

If you are familiar with the Bible, chances are you have opened this Holy Book on a random page at some point in your life. If so, you may have been surprised at how its contents somehow spoke the exact message you need to hear. I have delved into the Bible this way many times throughout my lifetime, and some of the most impactful discoveries I made occurred when I was a child. Throughout my ups and downs, the Bible has always stood as a wise friend and life guide, and it is as relevant today as it ever was. Technology is moving at breakneck speed, and in just two decades (since the smartphone boom), we have become a perpetually "connected" society... connected to a myriad of devices but sometimes disconnected from our faith.

The word of God is essential for those wishing to deepen their faith and feel close to Him. By using technology responsibly, you can harness myriad tools (from apps to podcasts) that will make it much easier to put your faith first, even in this busy, tech-savvy world. If you are enjoying your reading experience so far, I hope you can help other people of faith who feel lost in the elusive battle between faith and technology.

By leaving a review on Amazon, you'll show them that technology can help them strengthen their faith, build Godly character in a secular world, and deepen their prayer and meditation practices.

Thanks for your support. Keep reading to discover how you can transform doubts and spiritual struggles into opportunities for growth and advancement.

Scan the QR code below.

SCAN ME

CHAPTER SIX

DEEPENING PRAYER AND MEDITATION PRACTICES

Isaiah 55:8-9

"For my thoughts are not your thoughts, neither are your ways my ways,"
declares the Lord. "As the heavens are higher than the earth, so are my
ways higher than your ways and my thoughts than your thoughts."

In an age where our lives whirl with constant activity and our minds buzz from endless noise, finding a moment for genuine spiritual connection can sometimes feel like trying to hear a whisper in a storm. How often have you longed to quiet your spirit and deeply connect with God, only to find your thoughts scattered or your attention pulled away by the demands of daily life? This chapter invites you to explore a more profound, intentional way of engaging with Scripture and prayer, enriching your spiritual life through the ancient practice of scriptural meditation. This discipline offers peace and depth in our fast-paced world.

SCRIPTURAL MEDITATION FOR THE MODERN CHRISTIAN

Acts 17:28

"For in him we live and move and have our being. As some of your own poets have said, 'We are his offspring.'"

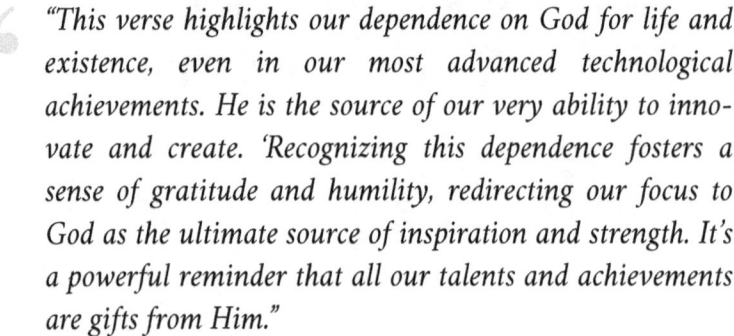

"This verse highlights our dependence on God for life and existence, even in our most advanced technological achievements. He is the source of our very ability to innovate and create. 'Recognizing this dependence fosters a sense of gratitude and humility, redirecting our focus to God as the ultimate source of inspiration and strength. It's a powerful reminder that all our talents and achievements are gifts from Him."

— PASTOR EMILY DAVID

Understanding Scriptural Meditation

Scriptural meditation is a form of prayer and reflection that involves a deep, contemplative focus on the Word of God. Unlike some forms of meditation that aim to empty the mind, scriptural meditation seeks to fill it with divine truth, utilizing Scripture as the focal point. This practice is about more than just reading the Bible; it's about allowing the Holy Scriptures to read us, penetrate our hearts and minds, and transform us from the inside out.

At its core, scriptural meditation is a dialogue with God. It's an active engagement where you speak to God through prayer and listen to what He has to say through His Word. This form of meditation can lead to a richer understanding of Scripture, as it encourages you to slow down and ponder God's words deeply, allowing the Holy Spirit to highlight truths and reveal insights that could be missed in a more cursory reading.

Techniques for Effective Scriptural Meditation

One highly fruitful method for scriptural meditation is lectio divina, a Latin term that means "divine reading." This ancient practice involves several steps that facilitate a deeper engagement with Scripture. Here's how you can practice lectio divina:

1. **Read**: Begin by slowly reading a passage of Scripture. Pay attention to any word or phrase that stands out to you.
2. **Meditate**: Reflect on the text. Think about what God might be saying through the highlighted words or phrases. How do they resonate with your current life circumstances?
3. **Pray**: Respond to the insights you've received. Speak to God about the passage and what it has stirred in you. Ask questions, seek clarification, or offer praise based on what you've learned.
4. **Contemplate**: Rest in God's presence. This is a moment of silent worship and adoration, where you let go of your thoughts and enjoy being with God.

Incorporating Context and Commentary

Understanding your study passages' historical and cultural context can enrich your scriptural meditation. This background can provide significant insights into the text, making your meditation more informed and nuanced. For instance, knowing the cultural significance of shepherding in ancient Israel could deepen your understanding of Jesus's role as the Good Shepherd.

Consider using reputable Bible commentaries or study guides that offer explanations and interpretations of Scripture from trusted theologians and scholars. These resources can be precious when dealing with complex passages or those that have been widely debated. However, always ensure that such tools supplement rather than replace the primary activity of engaging directly with Scripture.

Setting Goals for Regular Practice

Integrating scriptural meditation into your daily routine requires intentionality. Start by setting realistic, achievable goals. Consider meditating on Scripture for ten minutes daily, gradually increasing the time as you become more comfortable with the practice. Choose a specific time and place that minimizes distractions— early morning at your kitchen table, during a lunch break at a quiet corner of your office, or in your living room after dinner.

To keep track of your meditation journey, consider keeping a meditation journal. After each session, jot down the passage you meditated on, any insights you gained, and how you felt during the process. This record helps you see your growth and can encourage you when meditation feels challenging or unfruitful.

By embracing scriptural meditation, you're not just learning about God—you're learning from Him. This practice can transform your spiritual life, providing a path to deeper understanding, peace, and connection with the divine. As you commit to engaging with God's Word through meditation, may you find the richness and depth that comes from dwelling in His presence and hearing His voice in the quiet moments of reflection.

DEVELOPING A CONSISTENT PRAYER ROUTINE

Prayer is the breath of the Christian life, the intimate dialogue that sustains your spiritual vitality. Think of it as your daily check-in with God, a time to voice your deepest concerns, express gratitude, and receive guidance for the path ahead. The benefits of maintaining such a consistent prayer routine extend beyond spiritual health, touching every aspect of your existence. Emotionally, regular prayer can be a source of comfort and peace, helping to alleviate stress and anxiety by casting your cares upon God, who cares deeply for you. Psychologically, it reinforces a sense of purpose and connection, grounding your identity in who you are in Christ rather than the fluctuating circumstances of life. Spiritually, it keeps you rooted and growing in your relationship with God, ensuring that your faith does not stagnate but continues to flourish even in challenging times.

To integrate this critical discipline into your life, it is essential to establish a prayer routine that resonates with your lifestyle and personal rhythms. For some, this may mean setting aside specific hours of the day for prayer—morning, noon, and night—following the example of Daniel, who, despite his demanding role in the Babylonian empire, consistently made time to pray three times a day (Daniel 6:10). This fixed-hour prayer can help structure your day, providing anchor points that remind you to pause and reconnect with God amid your daily tasks.

However, not everyone may find fixed-hour prayer feasible, especially with unpredictable schedules. In this case, embracing spontaneous prayer opportunities can be equally enriching. This could look like turning your commute into prayer time, using lunch breaks to step away and spend a few moments in a nearby park talking with God, or praying while doing household chores. What's important is not when or how long you pray but that you are making regular, deliberate efforts to communicate with God throughout your day.

Incorporating prayer prompts and journals can significantly enhance the depth and consistency of your prayer life. Prayer prompts are simple statements or questions that help focus your thoughts and open your heart to God's presence. For example, a morning prompt might be, "Lord, what part of my life needs your grace today?" or "Help me to see others as you see them." These prompts can serve as springboards into more profound prayer, guiding your thoughts and helping you engage more personally and profoundly with God.

Keeping a prayer journal is another powerful tool. It lets you record your prayers and track how they are answered over time. This practice provides a tangible reminder of God's faithfulness and responsiveness and enhances your prayer life by making it more reflective and intentional. You might start each journal entry with a date and a brief note about your current circumstances or feelings, followed by your prayer for the day. Reviewing previous entries can be incredibly encouraging, especially during times of doubt or struggle, as you recall how God has worked in your life.

Community prayer practices also play a direct role in enriching your personal prayer life. Engaging in prayer with others—whether in a church setting, a small group, or an online prayer circle—can provide a sense of support and shared spiritual commitment that is deeply encouraging to others. These communal prayer times allow you to experience your community's collective faith, learn from others' prayers, and feel uplifted by the knowledge that others are praying with and for you. Many find that praying in a community helps sustain their prayer life, giving them new perspectives and renewed motivation to continue developing their prayer practices.

By embracing these strategies, you can build a prayer routine that fits into your daily life and profoundly enriches it, drawing you closer to God and fortifying your faith daily. As you continue to explore and implement these practices, may your prayer life become ever more vibrant and fulfilling, a true reflection of your ongoing relationship with your Creator.

OVERCOMING DISTRACTIONS DURING PRAYER AND MEDITATION

In our bustling lives, distractions are as common as the air we breathe, and this is no less true when we try to sit down for a time of prayer or meditation. How often have you settled in to pray or meditate, only to find your mind wandering to the grocery list or your phone lighting up with notifications that pull you back into the whirlwind of daily life? Recognizing and managing these distractions is essential for a fruitful spiritual practice, and it begins by identifying the common culprits that disturb our peace.

Digital notifications are the most pervasive modern distraction. Our phones, designed to keep us connected, often feel like chains that pull us away from deeper communion with God. Then there's the simple noise of daily life—traffic sounds, conversations, household noise—all of which can interrupt quiet reflection. And, of course, our wandering thoughts—plans, worries, daydreams—can make it hard to focus on spiritual matters. Acknowledging these distractions is the first step in managing them, allowing us to create strategies to minimize their impact during our precious moments of prayer and meditation.

Creating a conducive environment for your spiritual practices can significantly reduce these distractions. Start by designating a specific space in your home as your prayer or meditation spot. This space doesn't need to be large or elaborate, but it should be set apart in some way that signals to your mind, "This is where I meet with God." Make this space comfortable and inviting, perhaps with a comfy chair, a small table for a Bible or journal, and maybe a candle or two. Importantly, ensure this area is away from high-traffic parts of your home where interruptions are likely.

In this dedicated space, use ambient sounds to your advantage. While silence is golden, sometimes background sounds like soft instrumental music or nature sounds can help drown out the jarring noises of the outside world. Numerous apps and websites offer sounds specifically designed to aid concentration and relaxation. These can be particularly useful if you live in a noisy neighborhood or if complete silence feels too stark.

Mindfulness techniques are invaluable tools for focusing the mind during prayer and meditation. A straightforward technique is focused breathing, which involves taking slow, deep breaths and paying attention to the sensation of the air moving in and out of your body. This practice helps quiet the mind and physically relaxes the body, making it easier to sit still and concentrate. Another technique is visualization, particularly visualizing scriptural themes. For instance, while meditating on Psalm 23, you might visualize green pastures and still waters, imagining yourself in the serene landscape as God restores your soul. These techniques anchor your mind, drawing it away from distractions and back to the focus of your meditation.

Dealing with intrusive thoughts is another challenge many people face in spiritual practice. It's important to remember that such thoughts are a normal part of the human experience, not a sign of failure in your prayer or meditation efforts. When you notice your thoughts wandering, gently guide them to your prayer or Scripture focus without self-criticism. Some find it helpful to imagine placing their distracting thoughts on a leaf and watching it float away down a stream—acknowledged but set aside.

Another practical strategy is to use a "worry pad." Before you begin your prayer or meditation, take a moment to jot down any pressing thoughts or concerns on a notepad. This act can help clear your mind, reassuring it that those thoughts are held somewhere safe and can be revisited later. This frees you up to focus on your spiritual practice without fear of forgetting something important.

By integrating these strategies—creating a dedicated space, using ambient sounds, practicing mindfulness techniques, and managing intrusive thoughts—you can significantly enhance the quality and depth of your prayer and meditation. This intentional approach minimizes distractions and enriches your spiritual journey, helping you connect with God on a deeper level. As you refine your practice, may you find greater peace, focus, and spiritual insight in your moments of quiet communion with the Divine.

USING TECHNOLOGY TO AID PRAYER AND MEDITATION

2 Corinthians 9:8
"And God can bless you abundantly, so that in all things at all times, having all that you need, you will abound in every good work."

In this digital era, when technology infiltrates every aspect of our lives, it's worth exploring how it can be leveraged to enrich our spiritual practices of prayer and meditation. While there's a justified caution about the distractions technology can introduce, it can be a powerful ally in deepening your connection with Him when used wisely.

Consider the various apps and digital resources designed to enhance prayer and meditation. These tools offer everything from guided prayer meditations that walk you through various prayer models to digital rosaries that help you keep track of your prayers. Scripture-based meditation apps provide daily verses with thoughtful reflections and prompts, making it easier to engage with God's Word regularly. For instance, apps like Pray As You GO combine Scripture readings with music and questions for reflection, turning a simple daily commute into a rejuvenating connection time with God. Similarly, Abide offers guided Christian meditations, helping you focus, de-stress, and sleep better, all through Biblical truth.

The key to these digital tools is in how they're used. They should complement, not replace, the personal elements of your prayer life. These apps can help structure your meditation sessions, providing a guided framework that can be especially helpful for those new to meditation or those struggling to focus during prayer. The content provided, often crafted by theologians and seasoned prayer leaders, can bring new insights and depth to your understanding and practice.

Beyond apps, technology reminds us of prayer needs or scheduled meditation times. Most smartphones allow you to set customizable alerts, which can be used to remind you of specific prayer requests or prompt you to take time to meditate. Setting up a prayer alert for midday can be a great reminder to pause, refocus on God, and spend a few minutes praying, regardless of how hectic your day is. These little nudges help integrate prayer into the rhythm of daily life, making it a regular practice rather than an afterthought.

Moreover, the online world offers a wealth of workshops and webinars that delve into advanced techniques in prayer and meditation. These sessions can be invaluable for personal growth, offering teachings from experienced spiritual leaders that might not be otherwise accessible. Platforms like Zoom or Google Meet host live interactive sessions where you can learn from and even ask questions directly to these leaders, making it a dynamic learning environment. These online workshops broaden your understanding and connect you with a community of believers equally eager to deepen their spiritual practices.

However, as beneficial as these tools can be, it's all about maintaining balance. The convenience of digital aids must not detract from the essence of prayer and meditation, which is to foster a more profound, personal connection with the Lord. It's easy to let the passive consumption of digital content replace active, heartfelt engagement in prayer and meditation. To prevent this, use technology as a starting point or an enhancement rather than the entirety of your spiritual practice. Let these tools spark your thoughts and open your heart, but spend time in silence, listening for God's voice beyond the screen.

By thoughtfully integrating technology into your prayer and meditation practices, you can enhance your spiritual discipline while maintaining the personal depth and connection these practices are meant to cultivate. As we traverse the digital age, let us use every tool available to draw closer to God, letting technology serve our spiritual growth rather than distract us from it. (Psalm 119:105)

As we conclude this exploration of deepening prayer and meditation practices through technology, remember the overarching goal —to foster a more prosperous, consistent, and reflective spiritual life. With intention and care, technology can significantly support this goal, helping us remain connected to God amid our busy, tech-saturated lives. As we move into the next chapter, we'll shift our focus from our prayers to extending this enriched spiritual life outward, engaging the world around us with the love and truth we've cultivated through these disciplined practices.

CHAPTER SEVEN

ADDRESSING DOUBTS AND SPIRITUAL STRUGGLES

Colossians 3:23

"Whatever you do, work at it with all your heart, as working for the Lord, not for human masters."

Have you ever questioned the beliefs that once seemed so solid and unshakeable? Perhaps you've wrestled with doubts that crept in during quiet moments, or maybe they've surged forth in times of trial, leaving you to wonder about the nature of faith itself. You're not alone in this. Every believer, at some point, encounters moments of doubt. These aren't just the shadows on our spiritual walk; they can be the avenues through which our faith deepens and matures.

EMBRACING DOUBTS AS A PATH TO DEEPER FAITH

Normalizing Spiritual Doubts

Experiencing doubt does not make anyone a lesser disciple of Christ. Some of the most revered figures in the Bible grappled with uncertainty and questioning. Take Thomas, often dubbed Doubting Thomas, who couldn't believe the resurrection of Jesus until he saw Him with his own eyes. His doubts didn't disqualify him; they set the stage for a profound declaration of faith once he encountered the risen Christ. Or consider Elijah, who, despite witnessing God's immense power at Mount Carmel, found himself beset by doubts and fears because of Jezebel's threats, fleeing into the wilderness in despair.

These stories are more than ancient narratives; they mirror our experiences with doubt. They remind us that spiritual doubts are part of the human condition, a common thread in the fabric of faith. Acknowledging this can relieve the heavy burden of thinking we must always possess unwavering certainty. By seeing doubt as a shared aspect of faith, we can more freely discuss and address our uncertainties without fear of judgment, fostering a more supportive and understanding faith community.

Constructive Responses to Doubts

When doubts arise, the response matters immensely. One practical approach is to seek knowledge actively. This doesn't just mean a superficial search but a deep, earnest study of Scripture and Christian doctrine. Engaging with the Bible, theological books, and even discussions with knowledgeable leaders can illuminate confusing topics and provide reassurance.

Honest prayer is another vital response. This means coming before God with all your uncertainties and asking Him to guide you. It's about being transparent in your communication with God, not holding back for fear that your questions are too trivial or your doubts too great. Remember, God desires a relationship with us that is built on authenticity, not pretense.

Furthermore, discussing your doubts with trusted spiritual mentors can be incredibly helpful. These are individuals who not only understand the complexities of faith but also care about your spiritual well-being. They can offer biblical insights, share personal experiences of navigating similar doubts, and provide the kind of perspective and encouragement that's hard to find elsewhere.

Doubts as Opportunities for Growth

Rather than viewing doubts as stumbling blocks, we can see them as stepping stones to deepen our faith in Jesus. This shift in perspective is profound. It allows us to approach our doubts with curiosity and openness rather than fear and avoidance. When we explore our doubts, we engage in a process of discovery that can lead to more robust, more resilient faith. It's understanding that faith isn't the absence of doubt but the means to seek truth amid uncertainty. It is the substance of things we hope for and the evidence of things we have not yet seen (Hebrew 11:1)

Testimonies of Overcoming Doubt

Real-life stories of believers who have wrestled with and overcome their doubts can be incredibly inspiring. For instance, consider the story of Sarah, a college student who began questioning her faith after encountering various worldviews and philosophies in her studies. Her journey through doubt involved seeking answers through apologetics, engaging in honest prayer, and finding support from a campus Christian group. These steps not only helped her to reaffirm her faith but also equipped her to help others who might be facing similar questions.

Then there's Michael, a businessman who doubted God's goodness during a particularly tough economic period that threatened his company. Through pastoral counseling, personal study, and the unwavering support of his church community, he found a way through his doubts, emerging with an intact faith that was more nuanced and empathetic toward others in crisis.

These stories underscore the transformative power of confronting and engaging with our doubts. They remind us that while the road may be rough, the journey through doubt can lead to a faith that is not only maintained but magnified in its depth and sincerity. As you step out in faith, let these testimonies inspire you to seek, question, and grow. Embrace this aspect of your spiritual journey, knowing that on the other side of doubt often lies a stronger, more vibrant faith, ready to stand firm in the face of life's uncertainties.

SPIRITUAL DRY SPELLS: CAUSES AND CURES

Psalm 119:105

"Your word is a lamp for my feet, a light on my path."

Have you ever felt like you're walking through a spiritual desert, where prayers seem to echo into emptiness and the vibrant faith you once held feels distant and dry? You're not alone in this experience. Many believers grapple with "spiritual dry spells"—seasons where God feels distant, and our spiritual practices seem fruitless.

Understanding the causes of these dry periods can be the first step toward renewal.

One common trigger for spiritual dry spells is burnout, which can occur when we overextend ourselves in work, ministry, or daily responsibilities, leaving little energy for personal spiritual nourishment. Life transitions such as moving to a new city, changing jobs, or significant changes in family dynamics can also disrupt our spiritual routines, making it challenging to maintain a connection with God. Unconfessed sin is another profound but often overlooked cause; it can create a barrier between us and God, subtly eroding our peace and spiritual vitality. Lastly, a lack of community can intensify feelings of spiritual isolation. We are designed for fellowship; without it, our spiritual lives can feel barren.

Certain spiritual practices can act as rain upon the parched soil of our souls during these dry spells. Consider the refreshing nature of spiritual retreats; these intentional times away from daily routines can provide the space and silence necessary to reconnect with God. Whether it's a weekend away at a retreat center or simply a day set aside for extended prayer and reflection at home, retreats allow us to focus solely on our spiritual lives without the usual distractions.

Fasting is another powerful practice during dry spells. By voluntarily abstaining from certain foods or activities, we can sharpen our spiritual sensitivity, reminding ourselves that our sustenance comes from bread alone and every word from God's mouth. This discipline can bring clarity and renewal as it shifts our dependence away from the material and toward the spiritual.

Renewing commitment to daily devotions can also play a role in watering our dry spiritual ground. This could mean revisiting neglected practices such as daily prayer or Scripture reading or incorporating new forms of devotion like meditative reading or journaling. These practices help to lay a consistent foundation, allowing God's truth to seep slowly back into the crevices of our hearts.

Artistic expressions of worship, such as painting, writing, or making music, can also be particularly potent during dry spells. These creative outlets offer unique ways to engage with God's beauty and truth, often bypassing intellectual barriers to touch the heart directly. Engaging in worship through art can be a profoundly personal way to reignite passion and connection to the divine.

Patience and perseverance are essential companions as we walk through spiritual dry spells. It's important to remember that these seasons are just that—seasons. They do not last forever, even though it can be challenging while we endure. Maintaining spiritual disciplines, even when they seem fruitless, acts as a statement of faith, affirming our belief in God's presence and activity, even when we don't feel it. It's akin to planting seeds in winter, trusting that spring will come.

During these times, Scripture offers profound encouragement. Verses like Psalm 42:5, where the psalmist speaks to his soul, saying, "Why are you downcast, O my soul? Why so disturbed within me? Put your hope in God, for I will yet praise him, my Savior and my God," remind us that experiencing spiritual lows is expected. Moreover, they encourage us to continue hoping and praising, even in the desert. Isaiah 43:19 also brings comfort with God's promise: "See, I am doing a new thing! Now it springs up; do you not perceive it? I am making a way in the wilderness and streams in the wasteland." This reassurance of renewal and divine intervention can be a lifeline in moments of spiritual desolation.

Understanding the causes of our spiritual dry spells and actively engaging in practices that foster renewal can help us get through these challenging seasons with hope. Remembering, even the driest desert can bloom again under the restoring rains of God's grace and care.

RESPONDING TO LIFE'S UNANSWERED PRAYERS

Have you ever found yourself on your knees, eyes closed, heart open, pouring out your deepest needs to God, only to feel your prayers echo into a void? Unanswered prayers can be some of the most spiritually challenging experiences for believers. They test not just our faith but our understanding of how prayer works. It's essential to explore what the Bible tells us about why some prayers may not be answered in the ways we anticipate and how we can grow from these experiences.

Understanding Unanswered Prayers

The complexity surrounding unanswered prayers often concerns the clash between our expectations and God's intentions. In Scripture, we see countless examples where God's responses to prayers were not immediate or did not align with the desires of the prayerful. This discrepancy isn't because God is indifferent or not listening; it's rooted in His omniscience and sovereignty. God sees the entire tapestry of our lives—past, present, and future—while we see only the threads nearest us. This divine perspective means He may withhold an answer or respond differently, knowing that it ultimately serves a greater good or spurs essential growth in us.

For instance, Paul's repeated prayer for removing the "thorn in his flesh" is detailed in 2 Corinthians 12:7-9. God's answer was not the removal of the thorn but a reassurance of His grace, which

Paul found sufficient. This response shifted Paul's understanding of strength and power, teaching him that his weakness was a vessel for divine strength. Here, the unanswered prayer led to a profound spiritual revelation that strength comes from dependency on God, not self-sufficiency.

Reframing Expectations

Reframing our expectations of prayer is essential to navigating unanswered prayers. Prayer is not a transactional dialogue where requests are met with immediate fulfillment. Instead, it's a relational dialogue to align our will with God's will. It's about transforming our desires to fit better what God knows we truly need rather than what we believe we need. This perspective requires a shift from treating prayer as a cosmic wish list to embracing it as a dynamic communion with God, where praying shapes us just as profoundly as the answers we receive.

Encouraging this shift involves regular, honest reflection on the nature of our prayers. Are we seeking God's will or merely asking Him to endorse ours? This introspection can lead to a more mature prayer life that cherishes the relationship and conversation with God above the outcomes of our requests.

Maintaining Faith in the Silence

Maintaining faith can feel like holding on to smoke when God seems silent—elusive and disheartening. Yet, these moments of divine silence are not voids of neglect but periods of profound spiritual deepening. They are invitations to trust God's timing and plan, to remain steadfast in our spiritual disciplines, and to embrace the mystery of God's sovereign plans. Continuing practices such as daily Scripture reading, meditation, and communal worship during these times to keep our spirit open, press on in our faith, and sustain us until we receive clarity or peace.

Incorporating practices such as silent contemplation can also be transformative. It allows us to be still and know that He is God (Psalm 46:10), to listen rather than speak, and to find peace in His presence—even when answers aren't forthcoming. This discipline helps us detach our well-being from the reception of immediate answers and anchors us in the reality of God's unchanging nature.

Learning From Biblical Examples

Biblical figures like Job, David, and Jesus provide profound insights into dealing with unanswered prayers. Job's story is particularly poignant. Despite severe losses and immense suffering, with no answers to his desperate questions, Job remained rooted in his integrity and faith in God's justice. His faithfulness to God, despite unanswered prayers, led to eventual restoration and blessings beyond his earlier state.

In the Psalms, David often expressed deep despair and confusion over what he perceived as God's silence. Yet, he consistently circled back to a foundational trust in God's faithfulness, using the act of lament as a bridge back to a reaffirmation of his faith.

Even Christ, in the garden of Gethsemane, faced the silence of unanswered prayer when He asked for the cup of suffering to pass from Him. His ultimate submission to the Father's will, "not as I will, but as You will" (Matthew 26:39), exemplifies this profound trust humans are called to emulate.

These examples are not just historical accounts but mirrors reflecting our struggles with unanswered prayers. They teach us resilience, the depth of trust, and the peace from surrendering to God's grander plan. As you face your unanswered prayers, let these stories remind you that you are in good company. The path is trodden by the footprints of saints who learned that the true answer often isn't in the outcome but in the transformation that occurs through times of waiting, trusting, and believing in the One who promises to make all things work together for our good to those who love Him and are called according to His purposes and glory (Romans 8:28).

THE ROLE OF THE CHURCH IN SPIRITUAL TRIALS

Proverbs 4:7
"Wisdom is the principal thing; therefore, get wisdom: and with all thy getting – get understanding, too."

When the winds of spiritual trials blow hard against our faith, the church can be a beacon of hope and community support. Perhaps more than any other time, it's in these times that the church's role as a nurturer and protector of its flock becomes most evident. Engaging deeply within the church community during these periods can provide solace and tangible support to guide you through the storm.

The church is designed to be more than a place of worship; it is a community where believers can find encouragement, understanding, and practical support during tough times. This support manifests in various ways. For instance, many churches offer specific ministries aimed at helping members deal with grief, addiction, or family crises. These ministries often provide resources such as counseling, support groups, and practical assistance like meals or childcare, which can be invaluable when struggling to keep your spiritual footing. Engaging with these resources allows you to experience the tangible love and support of the body of Christ, reminding you that you are not alone in your struggles.

Furthermore, the church community offers a unique kind of empathy and encouragement. Sharing your burdens with fellow believers committed to walking alongside you can lighten the emotional load and provide new perspectives on your trials. This communal support is rooted in a shared understanding of faith and the recognition that everyone, at one point or another, faces challenges that test one's faith. By leaning into this supportive network, the church family's collective strength and prayers can uplift and sustain you.

Pastoral Care and Counseling

In times of deep spiritual struggle, the guidance and support of pastoral care can be a lifeline. Pastors and spiritual leaders are often equipped with theological training and insights into counseling that can help you discern the complexities of your spiritual trials. Seeking pastoral care can provide a safe space to express your doubts, fears, and struggles without judgment. Here, you can explore your feelings and questions under the guidance of someone who can offer biblically grounded counsel and encouragement.

Moreover, many pastors are trained to recognize when congregants need more specialized help than they can provide, such as therapy from a licensed professional. In such cases, they can assist you in finding the right resources and support, ensuring that your spiritual and emotional needs are addressed comprehensively. This level of pastoral care is invaluable, as it ensures that all aspects of your well-being are considered and cared for, which is necessary for healing and growth during spiritual trials.

Group Studies and Prayer Groups

Participating in small group studies or prayer groups can significantly impact our spiritual health, especially during difficult times. These groups provide a sense of belonging and community, vital for anyone navigating the waters of doubt or despair. The focused nature of these groups—whether it's a Bible study group delving into the Scriptures or a prayer group meeting to share and lift concerns—creates an environment where you can explore your faith in an intimate, supportive setting.

These groups often become sources of mutual support, where members encourage one another and share their spiritual journeys. Studying the Bible together can also bring new insights and revelations that may be missed in solitary study, helping to strengthen your faith. Similarly, prayer groups offer a powerful communal prayer experience that can reinforce your personal prayer life. The collective faith and intercession can be profoundly comforting, reminding you of God's presence and the power of prayer.

Service and Outreach Opportunities

Engaging in service and outreach activities can offer a renewed sense of purpose and perspective that is healing during spiritual doubt or dry spells. Serving others allows you to step outside your struggles and contribute to the welfare of others, which can shift your focus from your problems to the needs of those around you. This shift in perspective can be incredibly therapeutic and uplifting.

Moreover, acts of service can reinforce Christ's teachings about love and charity, bringing the Scriptures to life in practical ways. Whether helping at a food bank, visiting the elderly, or participating in community clean-up events, these acts of service embody the love of Christ. They positively impact the lives of others and bring a sense of fulfillment and joy to the giver, which can be a powerful antidote to spiritual malaise.

By actively participating in these church activities, you allow yourself to be both a receiver and a giver of God's love, which can profoundly affect who you are and your vitality. Giving and a supportive church environment can restore your faith and reinvigorate your spiritual life.

As we close this chapter on addressing doubts and spiritual struggles, remember that the journey through these challenges is a part of the larger story concerning your faith. Each trial is an opportunity for growth, each question a step toward more profound understanding, and each moment of despair a chance to be lifted by the community of believers surrounding you. With its rich community support resources, pastoral care, group studies, and service opportunities, the church is ready to assist people in navigating life trials. Let this community be your strength, refuge, and guide as you explore, question, and grow in your faith in God. As we move forward, let us carry the lessons learned and the strength gained from our trials, using them to enrich our lives and those of others.

CHAPTER EIGHT

FAITH AND RELATIONSHIPS

I n the tapestry of life woven with divine threads, how do we ensure that our most intimate relationships reflect the beauty and strength of God's design? Imagine your relationships as gardens—spaces of potential and growth, which, when nurtured with the principles of God's Word, can flourish into havens of love, support, and spiritual vitality. This chapter delves into the profound connection between our faith and relationships, mainly focusing on the sacred bond of marriage.

MARRIAGE: A COVENANT UNDER GOD

In the biblical sense, marriage transcends the legal contract that many perceive to be; it is a covenant that mirrors the steadfast love and commitment between Christ and the Church. Ephesians 5:25-33 vividly depicts this divine parallel, urging husbands to love their wives as Christ loved the church—sacrificially and unconditionally. This Scripture elevates the understanding of marital love and sets a high standard for the mutual commitment required in this sacred union.

Understanding marriage as a covenant implies that it is not merely an agreement between two parties but a sacred promise involving the couple and God Himself. This triadic relationship underscores the seriousness and sanctity of the marriage vow, which goes beyond human agreements and taps into divine strength. The covenantal nature of marriage calls for an enduring commitment, one that does not waver when circumstances change but persists, driven by the grace and power provided by God.

Roles and Responsibilities

In a covenant, each partner has roles and responsibilities that contribute to the strength and health of the agreed relationship. Scripture provides clear insights into the roles spouses are to play. While these roles are complementary, they are grounded in mutual respect, love, and support. For instance, while husbands are called to lead and love sacrificially, wives are urged to respect and support their husbands. However, these roles are not about hierarchy but about harmony. They are designed to work in tandem, reflecting selfless love and mutual submission to one another, as highlighted in Ephesians 5:21, "submitting to one another out of reverence for Christ."

This balance of roles in the light of contemporary society can be challenging, but embracing this with a heart of love and service transforms marital relationships. It shifts the focus from individual desires to the health and well-being of the marriage union and, ultimately, the family. When executed in the spirit of love and respect, these roles foster a nurturing environment where both spouses can grow individually and together.

Communication and Spiritual Intimacy

Effective communication and deep spiritual intimacy with God are at the core of a thriving marriage. These elements act like water and sunlight to the garden of marriage—essential for growth and vitality. Effective communication involves more than just talking; it's about sharing the thoughts of the heart, listening empathetically, and understanding each other's deepest fears, hopes, and dreams. It also involves discussing the more practical aspects of life together, like managing finances or parenting, with honesty and openness.

To build spiritual intimacy, couples are encouraged to engage in shared prayer times, joint Bible study sessions, and open discussions about their faith journeys. These shared spiritual disciplines bring couples closer to each other and draw them nearer to God, creating a solid threefold cord that is not quickly broken (Ecclesiastes 4:12).

Navigating Challenges with Faith

No marriage is without challenges, but the covenantal perspective provides tools to understand these difficulties. Financial stress, differences in faith levels, and external family pressures can strain a marriage. However, approaching these challenges with faith and reliance on God's wisdom can turn potential breaking points into opportunities for growth and deepening commitment.

When financial troubles arise, for instance, using principles from Scripture regarding stewardship and trust in God's provision can guide couples in managing their resources wisely and peacefully. In differing faith levels, maintaining respect and understanding can foster spiritual growth and harmony. For external family pressures, setting healthy boundaries, guided by biblical principles, can protect the marital relationship while still honoring extended family relationships.

By integrating these biblical principles into the very fabric of marital life, you can cultivate a relationship that not only withstands life's challenges but also thrives, reflecting the beauty and permanence of God's covenant with His people. As you and your spouse grow together in love and faith, your marriage can testify to the transformative power of living out God's design for this sacred union.

PARENTING WITH GRACE AND TRUTH

Raising children in the faith is a profound privilege and a significant responsibility. As parents, you're not just nurturing bodies and minds; you're also shepherding hearts. This dual role demands a balance of grace and truth, mirroring Christ's interaction with us. Grace without truth can lead to permissiveness that overlooks the need for moral boundaries, while truth without grace can foster a legalistic environment that may stifle understanding and love. Striking this balance means embedding your parenting in the foundational principles of Christian parenting, where passion, discipline, instruction, and spiritual leadership converge to guide your children toward a life rooted in Christ.

To parent with grace means to show unmerited kindness and patience, reflecting the unconditional love that God shows us. This doesn't mean avoiding discipline; it involves discipline that underscores love and correction rather than punishment. Proverbs 22:6 teaches us to "train up a child in the way he should go: and when he is old, he will not depart from it." This training involves setting clear expectations and consistent boundaries, communicating with love, and reinforcing fairness. Ephesians 6:4 complements this by urging parents not to provoke their children to wrath but to bring them up in the nurture and admonition of the Lord. Nurturing their growth through loving discipline teaches them about consequences and responsibility while always affirming their value and worth.

In addition to discipline and instruction, spiritual leadership is paramount. You're called to be the first and most influential example of godly living to your children. This doesn't mean you must be perfect; it means striving to model behaviors like forgiveness, kindness, and faithfulness. Establishing regular family devotions can be a powerful way to set this tone. These can be simple moments each day or week where you gather to read Scripture, discuss its application, and pray together. It's making faith a natural and consistent part of everyday life, not only addressed within a church's walls.

Moreover, serving together as a family in church activities or community service projects can profoundly impact your children's understanding of faith in action. It teaches them about selflessness, the joy of helping others, and the importance of community. These shared experiences can also strengthen your family's bond and provide practical contexts for discussing and living out the values you teach at home.

Handling tough questions and doubts with openness and honesty is also human. As children grow, they will inevitably question the world around them, including their faith. Addressing these questions with truth and patience is vital. It shows your children that their thoughts and doubts are valid and that their faith journey is a personal discovery and growth, not merely acceptance of secondhand convictions. For instance, if your child is struggling with why God allows suffering, discuss it openly. Explore Scripture together, pray for wisdom, and consult additional resources like books or videos from trusted Christian leaders. This approach helps answer their immediate questions and equips them with tools to explore their faith independently.

Integrating these practices into parenting creates an environment where grace and truth are taught and tangibly experienced. This environment helps children grow into adults who not only know about Jesus but also know Him personally and reflect His love in their lives. As you continue to guide your children through the hardships of life and faith, remember that every effort you make plants seeds in them that, with prayer and perseverance, will grow into a harvest of righteousness in their lives.

FRIENDSHIPS THAT FOSTER FAITH

In the mosaic of our spiritual lives, friendships represent colorful, vital pieces that enhance the beauty and strength of our faith journey. Proverbs 13:20 teaches us, "Walk with the wise and become wise, for a companion of fools suffers harm." This wisdom underscores the importance of surrounding ourselves with individuals who share, encourage, and strengthen our faith. The choice of friends, therefore, isn't just a social preference but a spiritual strategy. Friends who embody and promote godly principles can elevate our commitment to God, challenging us to grow deeper in our faith and fully live out our convictions.

When choosing friends, look for qualities that reflect a heart aligned with God's values—integrity, kindness, patience, and a commitment to the truth are essential. These characteristics are foundational for friendships that endure life's storms and foster mutual spiritual growth. Such friends serve as iron sharpening iron, as described in Proverbs 27:17, where each interaction can refine and enhance each other's faith. In practical terms, this may mean choosing friends who are unafraid to challenge you when you stray from your values or encourage you when your spirit wanes. They are the ones who pray for you fervently and stand by you unconditionally.

Building and maintaining these godly friendships often requires intentional actions. Regularly participating in shared spiritual activities such as Bible studies, prayer meetings, and church events can significantly strengthen these relationships. These activities allow friends to delve into God's Word together, confronting and discussing spiritual truths and challenges, which can lead to deeper understanding and connection. For example, a weekly Bible study group provides a structured opportunity for friends to discuss Scripture, apply its lessons to their lives, and support each other in those applications. Similarly, attending church services together can reinforce shared beliefs and offer regular touchpoints for spiritual encouragement and accountability.

The role of friendships in providing spiritual support cannot be overstated, especially during personal trials or spiritual dry spells. Godly friends are anchors when challenges arise, offering practical and spiritual support. They remind us of God's promises when our faith falters and stand with us in prayer when our strength fails. These relationships often become the tangible expression of God's love and care, reflecting His compassion and commitment through human actions. For instance, during a prolonged illness, a friend might pray for healing, provide meals, help with errands, or offer their presence as a comforting reminder of God's love.

However, not all relationships foster our spiritual growth; some might even hinder it. Recognizing and dealing with toxic relationships requires wisdom and discernment from the Holy Spirit to maintain spiritual health. A toxic relationship is characterized by consistent patterns of behavior that drain your energy, undermine your peace, or cause ongoing conflict. These relationships often lack the mutual respect and uplifting nature that characterize healthy friendships. Setting healthy boundaries in such relation-ships is not only wise but necessary. This may involve honest conversations about how certain behaviors affect you and requests for changes in the relationship dynamic. In cases where reconcilia-tion proves challenging or harmful behaviors persist, it may be necessary to distance yourself from the relationship. This isn't an act of unkindness but an essential step toward guarding your spir-itual well-being.

In managing these dynamics, it's essential to approach the situa-tion with grace and truth. Grace involves treating others with love and dignity, regardless of their actions. Truth is honesty about how their behavior impacts you and your spiritual life. This balance helps us discern the complex waters of confronting toxic behavior while embodying Christ's love and forgiveness. Whether through setting boundaries, seeking mediation, or, if necessary, taking time apart, handling toxic relationships with biblical wisdom is essential for maintaining both personal peace and spiri-tual integrity.

In cultivating friendships that foster faith, remember that these relationships are among the most significant investments you can make in your spiritual journey. They require effort, patience, and a heart open to giving and receiving godly love and wisdom. As you walk this path, let your friendships reflect the divine relationship —marked by love, sustained by truth, and aimed at mutual growth and glorification of God.

DEALING WITH CONFLICT IN A CHRIST-LIKE MANNER

In the walk of faith, few things test our commitment to Christ's teachings as profoundly as how we handle conflict. Conflict with our families, friends, or church communities is inevitable. Yet, the Bible offers us robust guidance on navigating these challenging waters with grace and integrity. Key passages like Matthew 18:15-17 and Romans 12:18 do not merely suggest but command us to pursue peace and reconciliation with a spirit of humility and forgiveness. These Scriptures are not just ancient texts but living words that can transform how we handle disputes in our everyday lives.

When conflict arises, our first call is to forgiveness. This doesn't mean dismissing the wrongdoing or the hurt caused but choosing a path of mercy over resentment. Forgiveness is the bedrock of conflict resolution in the Christian context because it reflects the grace we receive from God. It's about letting go of the desire for retribution and allowing God's peace to rule our hearts, as Colossians 3:13 urges us to bear with each other and forgive one another just as the Lord forgave us. This step is important because it frees us from anger and bitterness, allowing us to resolve conflicts with clarity and compassion.

Practical Steps for Resolving Conflicts

Navigating through conflict in a manner that honors God requires more than good intentions; it requires strategic, thoughtful actions. The first step is to pray for personal guidance, peace, and the other person's heart. Prayer positions our hearts correctly and prepares us for reconciliation. Next, approach the person privately, as Matthew 18:15 advises. This one-on-one discussion avoids unnecessary drama and shows respect for the relationship by not exposing the conflict to a broader audience prematurely.

In these conversations, clear and compassionate communication is critical. Start by expressing how much you value the relationship, then describe the issue as you see it, focusing on behaviors and impacts rather than attributing motives. Use "I" statements to express how the actions affected you, which can help in keeping the conversation non-accusatory. For instance, instead of saying, "You don't care about my feelings," try, "I felt hurt when my feelings were not considered." This approach helps in keeping the lines of communication open and constructive.

Listening is just as essential as speaking. Allow the other person to share their perspective fully. This doesn't just mean hearing them out but actively listening, showing empathy, and trying to understand their point of view. Often, conflicts escalate because of misunderstandings that could be cleared up simply by listening more attentively.

The Role of Mediation and Counseling

Sometimes, conflicts are too complex or emotionally charged to resolve without outside help. This is where mediators or counselors can be invaluable. They serve as neutral third parties who can help de-escalate tensions and facilitate a more objective perspective. Mediators, in particular, are trained to help conflicting parties find mutually acceptable resolutions, often through structured processes that ensure each person's concerns are heard and addressed.

On the other hand, Christian counseling can provide deeper explorations of the issues behind the conflict, often bringing biblical principles into the conversation to guide the resolution process. This can be especially helpful in disputes that involve deeper emotional wounds or spiritual issues, as counselors can help address these underlying concerns in a safe, faith-based environment.

Restoring Relationships After Conflict

They rebuild trust, and restoring a relationship after a conflict takes time and effort from all parties involved. Continual forgiveness, ongoing communication, and a renewed commitment to the relationship are essential. Establish new patterns of interaction based on mutual respect and transparency. For instance, regular check-ins on how each party feels about the resolution process can prevent old wounds from festering.

It's also important to celebrate progress. Acknowledging milestones in the reconciliation process can reinforce the value of the relationship and the efforts made to preserve it. This doesn't mean the memory of the conflict is erased, but it's overshadowed by the commitment to move forward with grace and understanding.

Navigating conflicts in a Christ-like manner not only resolves disputes but also deepens our spiritual maturity and mirrors God's reconciliation with us through Christ. As you apply these biblical principles and practical steps in your life, may your efforts lead to peace and restoration that reflect the love of Christ to all those around you, turning conflicts into testimonies of grace and redemption.

Transitioning to the Next Topic

As we wrap up our discussion on navigating conflicts with Christ-like grace, we prepare to explore another vital aspect of our spiritual journey—engaging with the broader world through our faith. Just as we are called to handle conflicts within our relationships with grace, we must enter the world as ambassadors of Christ's peace and reconciliation. In the next chapter, we will delve into how our faith sharpens our interactions with society and how we can witness God's love in a world desperately searching for it.

CHAPTER NINE

CHRISTIAN COMMUNITY AND FELLOWSHIP

Have you ever walked into a room and instantly felt like you belonged? That warm, welcoming sensation that envelops you isn't by chance—it's the result of intentional community-building efforts. In the context of our churches, cultivating such an environment is crucial, not just for the comfort it provides but for the spiritual growth and unity it fosters among members. This chapter dives into the practical and transformative aspects of creating and nurturing vibrant church communities that reflect the inclusivity and love of Christ.

BUILDING STRONG CHURCH COMMUNITIES

Cultivating a Welcoming Environment

Creating a welcoming atmosphere in the church isn't just about friendly greetings; it's about crafting an environment where new and existing members feel genuinely valued and included. Think about the first time you visited a church. What made you return?

Was it the smiling faces that greeted you at the door, or perhaps the sense of being noticed and appreciated during your visit? These experiences are fundamental to fostering a sense of belonging.

To enhance this welcoming culture, consider the entire experience of someone new to your church, from the parking lot to the pew. Practical tips include having a well-organized greeting team that not only smiles and shakes hands but also provides information about the church services and activities. Training greeters to actively recognize and engage with newcomers can make a significant difference. Follow-up strategies are equally important. A simple welcome packet or a follow-up call can go a long way in making newcomers feel cared for and seen.

Additionally, hosting inclusive activities catering to various interests and demographics within your church can strengthen the community. For instance, arranging mixers that bring together different age groups and cultural backgrounds can break down barriers and enrich fellowship. These gatherings can be simple post-service coffee hours or themed events encouraging interaction and conversation.

Role of Leadership in Community Building

The role of church leaders in fostering a community spirit cannot be overstated. Leaders set the tone for inclusivity and engagement within the church. By embodying a vision of an open, heartfelt community, leaders can inspire others to do the same. This leadership involves more than preaching about community; it includes being actively engaged in the lives of church members.

Leadership training should emphasize the importance of building a cohesive community and equip leaders with the skills to nurture this environment. This training may cover topics such as conflict resolution, active listening, and cultural competency, which are vital in managing a diverse congregation. Leaders trained in these areas can better model inclusive behavior, encouraging the congregation to embrace and celebrate diversity within the church community.

Integrating Diverse Demographics

The beauty of the church often lies in its diversity. Integrating diverse age groups, cultural backgrounds, and social statuses can present challenges, but the benefits of such a rich tapestry are immense. Strategies for productive integration include offering varied worship styles that respect different cultural expressions and planning activities that address the specific needs of various demographic groups within the church.

For instance, while younger members might appreciate a contemporary worship style, older members might feel more connected with traditional hymns. Balancing these preferences by hosting different services or alternating music styles can foster greater inclusivity. Additionally, creating targeted groups or ministries that cater to specific demographics such as youth, singles, married couples, and seniors can ensure that the needs and interests of all church members are being met.

Community Events and Activities

Community events are the glue that binds the church community together. These activities provide opportunities for deeper interaction and fellowship beyond Sunday services. Planning and promoting these events requires careful consideration to ensure they cater to the broad spectrum of the congregation.

Organizing church-wide retreats, for example, can be a powerful way to deepen bonds among members. These retreats can combine spiritual teaching with recreational activities, creating a relaxed environment for members to connect on a deeper level. Service projects, such as community clean-ups or food drives, benefit the wider community and unite church members in a shared mission, reinforcing the values of service and cooperation.

Clear communication is critical to organizing and promoting these activities. Platforms such as newsletters, social media, and church bulletins can ensure that all church members are informed and invited. Encouraging members to invite friends and family can also extend the warmth of your church community to others and potentially draw new members into the fold.

By focusing on these areas—creating a welcoming environment, emphasizing the role of leadership, integrating diverse demographics, and organizing inclusive events—your church can build a robust and vibrant community that not only attracts members but deeply engages them. This engagement is essential for fostering spiritual growth and extending the reach of your church's ministry into the community and beyond. As you implement these strategies, watch your church transform into a dynamic hub of activity and fellowship, a true reflection of the kingdom of God.

SMALL GROUPS: DEEPENING FAITH TOGETHER

Imagine you're part of a setting where everyone knows your spiritual struggles and triumphs, your faith questions spark deep discussions, and you meet weekly to study the Word and share. This is the essence of small group participation within the church —a microcosm of community that profoundly impacts your spiritual journey. Small groups offer a unique blend of personalization, intimacy, and mutual support that can dramatically deepen one's faith. They provide a platform for more personalized Bible study, where questions can be addressed in depth and insights can be shared freely. The close-knit nature of these groups fosters deeper relationships as members regularly share their lives, praying for each other's needs, celebrating joys, and comforting one another during hardships. This consistent interaction builds trust and intimacy that is hard to achieve in more extensive church settings.

Moreover, small groups enhance accountability in spiritual practices, which is fundamental for spiritual growth. It's easy to neglect personal Bible study or prayer when life gets busy, but knowing that your group is counting on you to participate and contribute can motivate you to stay disciplined in your spiritual practices. This accountability can also extend to moral support, where group members encourage one another to live out their faith in practical ways throughout the week.

However, starting and sustaining small groups in a church requires thoughtful planning and consistent effort. Choosing the right leaders is foundational—individuals who are not only spiritually mature but also have a genuine passion for community and a knack for facilitating discussion. These leaders are essential in setting the group's tone, guiding the study, and ensuring everyone feels valued and heard. Training these leaders can involve equipping them with skills in effective communication, conflict resolution, and group dynamics management, all essential for maintaining the health and vibrancy of the group.

Setting clear expectations right from the outset can also significantly influence the success of a small group. Members should clearly understand the group's purpose, what is expected of each participant, and the ground rules for discussions. This clarity helps prevent misunderstandings and ensures the group operates within a framework that promotes respect, openness, and mutual support. Maintaining engagement over time can be challenging but is essential for the group's vitality. Regularly introducing fresh study materials, rotating leadership roles, and incorporating social or service activities can keep the group dynamic and prevent it from becoming stagnant.

Another critical aspect is selecting the right curriculum and study materials. The choice of what to study should consider the group members' diverse needs and spiritual maturity levels. While some groups might benefit from book-of-the-Bible studies that provide a broad overview of Scripture, others may find topical studies on issues like marriage, parenting, or Christian living more applicable to their current life stages. In either case, the materials should foster knowledge and practical application of the Scriptures to daily life.

Dealing with group dynamics and challenges requires wisdom and sensitivity—particularly in managing different personalities and potential conflicts. Small groups are cross-sections of broader church life, meaning they can include various perspectives and temperaments. Leaders should be adept at facilitating discussions in a way that gives everyone a voice and keeps the conversation anchored to the topic. Conflicts, when they arise, should be addressed promptly and with a focus on reconciliation and forgiveness. Leaders might sometimes find it necessary to have one-on-one conversations with members to resolve deeper issues, ensuring that the group remains a safe and supportive space for all members.

In fostering these intimate, dynamic small group environments, churches can significantly enhance the faith experiences of their members, creating spaces where deeper spiritual roots can grow. These small groups become places where the theoretical aspects of faith meet the practical, everyday challenges of living it out, providing a rich soil for spiritual growth and maturity.

SERVING AND BEING SERVED: THE ROLE OF ALTRUISM IN COMMUNITY

When you roll up your sleeves and offer your time and talents to help others within your church and community, you're not just filling a need but actively embodying Christ's command to love and serve one another. This service, deeply rooted in the teachings of Jesus, creates a ripple effect of love and support that not only meets immediate needs but also builds a stronger, more cohesive community. Think about the parable of the Good Samaritan—here was a man who went out of his way to help a stranger. This story isn't just a lesson in helping others; it's a call to see service as an integral part of our Christian life.

Serving within your community can take many forms and should ideally match your unique gifts and interests. Perhaps you have a knack for teaching and could volunteer in your church's education programs, from Sunday school classes to adult Bible studies. Maybe you're skilled in a trade and could help with maintenance or building projects. Or perhaps you have a heart for older people and could spend time visiting nursing home residents. The possibilities are as diverse as the skills present in any congregation. The key is to match these opportunities with personal talents, ensuring that each member can contribute meaningfully, enhancing their sense of purpose and connection to the community.

Such a service's impact extends far beyond its immediate benefits. Engaging in service projects can lead to significant personal growth. It shifts the focus from self to others, a movement that can be profoundly transformative. Serving others can also help you develop new skills, build strong relationships, and improve mental and emotional health. From a communal standpoint, when members of a church come together to serve, it strengthens the bonds between them, creating a network of support and mutual care. This unity is essential for a healthy church community that can withstand challenges and is pivotal in fostering an environment where members feel they belong and are valued.

To cultivate a culture of service within your church, it's critical that both leaders and members actively promote and participate in service opportunities. Church leaders can set a powerful example by being involved in and passionate about various service projects. Furthermore, implementing recognition programs can be a way to acknowledge and appreciate the efforts of volunteers. Such recognition not only serves to thank individuals for their contributions but also encourages a continuing commitment to service among the congregation.

In addition to recognition, providing proper training for volunteers can ensure their service is fruitful and fulfilling. Whether training teachers to provide the best educational experience, coaching volunteers on interacting with the homeless community, or training in health and safety protocols for food distribution, equipped volunteers are more confident and experience greater satisfaction in their service roles. Moreover, integrating service into the very fabric of church life is essential. This integration can be facilitated by regularly scheduling service projects, incorporating discussions about service into sermon applications, and encouraging small groups to adopt service projects as part of their regular activities.

In essence, fostering a culture of service within a church not only helps meet the practical needs of the community and congregation but also deeply enriches the spiritual lives of those serving. As each member contributes their time and talents, they answer Christ's call to serve and craft a living testimony of God's love in action. Through such acts of service, churches can extend their influence beyond their walls, touching lives and transforming communities in tangible ways. As you consider your role within your church, remember that every act of service, no matter how small, plays a part in the larger story of God's love for His people. Each opportunity to serve is a chance to reflect Christ's love, build up the body of believers, and make a lasting impact on the world around you.

ONLINE CHURCH SERVICES: ADVANTAGES AND CHALLENGES

In today's digital era, the church has an unprecedented opportunity to extend its reach far beyond the traditional walls of worship through online services. These virtual gatherings are not just a supplementary option; for many, they are a primary way to connect with their faith community, especially for those homebound, living in remote areas, or even those seeking a spiritual home online from other parts of the world. The advantage here is clear—online church services can bridge geographical gaps, bringing the message of Christ to a global audience at the click of a button.

The inclusivity of online services is profound. They open the doors to anyone with internet access, offering a spiritual lifeline to those who might not otherwise have the opportunity to engage with a church community. This can be particularly life-changing for individuals in isolated areas or those with disabilities that make attending in-person services challenging. Moreover, for people exploring faith, online services provide an accessible platform to learn more about Christianity without the potential pressure of physically stepping into a church. This accessibility broadens the church's reach and deepens its impact, allowing it to minister to a diverse and widespread congregation.

However, maintaining engagement in this virtual setting can pose unique challenges. Without the physical presence of a congregation, how does one capture and sustain the attention of an online audience? Interactive elements become vital. Incorporating features such as live chat functions during the service allows members to interact with each other and church leaders, fostering community and participation. Follow-up discussions in smaller groups post-service can also help deepen the sermon's understanding and maintain personal connections. These strategies aim to replicate the interactive nature of in-person services, helping to actively engage the congregation and making the virtual experience more dynamic and individual.

Technical Considerations and Set-Up

Several technical aspects must be considered when ensuring that online services meet the community's needs. The quality of the streaming service can significantly affect the congregation's engagement. No one wants to connect to a service only to experience freezing screens and poor sound quality. Investing in reliable streaming software is brilliant. Platforms like Church Online Platform or mainstream services like YouTube Live offer robust solutions for live-streaming church services. Additionally, ensuring the church has a strong internet connection and backup options can mitigate technical issues that could disrupt the service.

Hardware considerations, including professional-grade cameras and microphones that ensure high-quality audio and visuals, are equally important. This might require some investment, but improving the quality of the service can significantly enhance the worship experience for online attendees. Training staff and volunteers to manage this equipment during services is also essential, ensuring that every aspect of the service runs smoothly from a technical standpoint.

Addressing the Challenges of Online Worship

Despite the benefits, online worship inherently lacks some elements of personal connection fundamental to in-person church services. The warmth of a handshake, the shared smiles, and the communal singing are not easily replicated in a virtual environment. This lack of personal touch can sometimes make online attendees feel like spectators rather than active participants in worship.

To mitigate these issues, churches must intentionally create opportunities for personal connection. This could involve setting up virtual coffee hours before or after services where people can discuss the sermon or share life updates. Encouraging members to turn on their cameras during these times can create a more personal and engaging experience. Additionally, having a pastoral team dedicated to online ministry can help reach out to care for the virtual congregation, ensuring they feel seen and heard.

As we navigate the waters of digital church services, the goal remains clear—to make church accessible to everyone, everywhere, without losing the essence of community and personal touch that characterizes the body of Christ. By leveraging technology wisely and intentionally, we can create online church experiences that reach across miles, touch hearts, and change lives, one digital interaction at a time.

As this chapter closes, we reflect on technology's transformative role in modern worship. Each aspect contributes significantly to the church's broader mission, from building vibrant community connections to enhancing personal engagement through small groups, serving within our communities, and expanding our reach through online platforms. As we turn the page, we continue to explore how these digital advancements intersect with traditional church practices, aiming to harmonize the old with the new in our continual quest to spread the Gospel effectively in this ever-evolving world.

CHAPTER TEN

THE ROLE OF FAITH IN OVERCOMING LIFE'S CHALLENGES

Philippians 4:6-7

"Do not be anxious about anything, but in every situation, by prayer and petition, with thanksgiving, present your requests to God. And the peace of God, which transcends all understanding, will guard your hearts and minds in Christ Jesus."

Have you ever wondered where your next meal would come from or how you would pay next month's rent? Economic uncertainty can strike anyone, and it often feels like a dark and unpredictable storm. How we respond to this storm, especially regarding our financial decisions and faith, can profoundly reflect and shape our relationship with God. In this chapter, we delve into the principles of financial faithfulness, exploring how trust in God's provision can guide us through periods of economic instability and uncertainty.

FINANCIAL FAITHFULNESS: TRUSTING GOD IN ECONOMIC UNCERTAINTY

Biblical Principles of Financial Stewardship

When it comes to managing money, the Bible is not silent. Jesus said in Matthew 6:33, which encourages us to "seek first the kingdom of God and his righteousness, and all these things will be added to you," reminding us that our financial decisions should not just be about balancing budgets but prioritizing our spiritual commitments. Similarly, in Philippians 4:19, the Apostle Paul reassures us, "And my God will supply every need of yours according to his riches in glory in Christ Jesus." These passages underscore a profound truth—our financial stewardship is deeply entwined with our trust in God.

This stewardship involves more than careful money management; it encompasses generosity and a reliance on God's provision that often challenges our worldly instincts. For instance, the concept of tithing—giving a portion of one's income to the church—can seem counterintuitive during financial hardship. However, it's a practice that reinforces our dependence on God, not just our financial resources. It's about making a tangible declaration that our security lies in God, not our bank accounts.

Faith During Financial Hardship

Consider Joseph in Egypt, as recounted in Genesis. He faced abundance amid extreme famine, yet his reliance on God's guidance led him to survive, thrive, and save many lives during a severe famine. His story is a powerful testament to how we can invest in any economic climate by aligning our financial strategies with divine wisdom.

This doesn't mean that financial prudence is thrown out the window. Joseph's strategy involved saving during the years of plenty in preparation for the famine. This kind of wise management—budgeting, avoiding debilitating debt, and planning for future needs—is not just practical; it's biblical. It reflects a mindset prepared for ups and downs, trusting God's provision through both.

Practical Financial Faithfulness Strategies

Practically, this means setting up a budget that includes regular savings and charitable giving, no matter how tight our finances seem. It's about recognizing that every dollar is a blessing and a responsibility. Avoiding debt wherever possible is also vital. Proverbs 22:7 warns that "the borrower is the slave of the lender," a reality that can become painfully evident in times of financial crisis. By living within our means and planning for the future, we can maintain a position of financial stability that allows us to respond to God's call—whether helping a neighbor in need or supporting a mission project.

Testimonies of Divine Provision

The stories of modern-day believers who have experienced God's miraculous provision can be incredibly encouraging. Consider a family who, after faithfully tithing even during unemployment, received an unexpected financial gift that covered their rent at just the right time. Or a single mother who, despite her limited income, found her resources stretched just enough to meet her family's needs. These testimonies aren't just heartwarming—they're powerful reminders of God's faithfulness and responsiveness to our faithfulness.

In sharing these stories, we see a pattern—when we place our trust in God and adhere to His principles of financial stewardship, He meets our needs in ways that defy human logic. It's a testament to the truth of Matthew 6:33, not just as a spiritual principle but as a practical promise. As you navigate your financial challenges, remember these stories and Scriptures. Let them bolster your faith and guide your financial decisions, turning economic uncertainty into an opportunity for spiritual growth and a testament to God's provision.

HEALTH CRISES AND SPIRITUAL RESPONSES

When health challenges strike, it's like being thrust into a storm where the winds of uncertainty and the waves of anxiety threaten to overwhelm us. The Scriptures anchor during these times, offering profound comfort and strength. Isaiah 41:10, for instance, is a powerful source of encouragement. "Fear not, for I am with you; be not dismayed, for I am your God; I will strengthen you, I will help you, I will uphold you with my righteous right hand." This verse isn't just a reassurance; it's a promise of God's presence and support, which can be incredibly comforting when facing health issues that make you feel vulnerable and frightened.

Drawing on Scriptures during health crises does more than comfort; it connects us to the truth of God's enduring presence and faithfulness. Another significant verse, James 5:14-15, guides the sick to seek prayer and anointing from the church's elders, emphasizing the community's role in healing. Calling upon elders for prayer highlights the biblical practice of communal faith in action—where the spiritual support of one's community plays a pivotal role in the healing process. These scriptural encouragements remind us that we are not alone in our struggles; we are part of a faith community that stands with us.

The Role of Prayer in Healing

Prayer during illness is not merely about seeking physical healing but is a profound dialogue that fosters spiritual strength and peace. It's a time to express our fears, hopes, and frustrations. Prayer becomes a cathartic release and a stabilizing force, providing peace that, as Philippians 4:7 describes, "transcends all understanding, guarding your hearts and your minds in Christ Jesus." In this divine communication, we find a spiritual fortitude that sustains us, regardless of the physical outcome.

This spiritual communion through prayer is necessary because it aligns our hearts with God's will, enabling us to tread the emotional and psychological upheavals often accompanying health crises. It's also a time when we can experience profound insights into the nature of God's love and compassion. For instance, through prayer, a believer may find a deep sense of peace and purpose in suffering, discovering that trials might serve a greater good or inspire others facing similar challenges.

Balancing Medical Treatment with Spiritual Faith

Navigating medical decisions can be daunting, especially when options are complex or outcomes are uncertain. Integrating faith with medical treatment involves recognizing that God has provided a means of healing through medical science. Luke, the author of the Gospel of Luke and the Acts of the Apostles, was a physician, and his writings reflect a deep respect for medical science and divine intervention. This duality underscores that pursuing medical treatments and trusting in God's power are not mutually exclusive but are complementary aspects of seeking to heal.

Making informed decisions about medical treatments involves prayerful consideration, seeking wisdom from God, and consulting knowledgeable healthcare professionals. It also means being open to how God can work through medical means. For instance, a new treatment or a referral to a specialist may be viewed as an answer to prayers for healing, a reminder that God often works His will through earthly means.

Community Support and Faith

The role of the church community is indispensable in supporting those undergoing health crises. Practical help, such as meals, transportation to medical appointments, or assistance with everyday chores, is invaluable. Additionally, the spiritual support through prayer groups and pastoral visits provides emotional and spiritual sustenance. For example, a church prayer group may gather to pray specifically for someone undergoing surgery, providing spiritual support and a tangible expression of the community's care and concern.

This communal support reflects the New Testament church, characterized by believers bearing one another's burdens. Such support comforts those in need and strengthens the community's bonds, fostering a more profound sense of unity and shared faith. It's a beautiful, practical manifestation of the scriptural call to "rejoice with those who rejoice; mourn with those who mourn" (Romans 12:15), living out our faith in ways that tangibly bless others in their time of need.

NAVIGATING LOSS AND GRIEF WITH HOPE

When the shadow of loss darkens our lives, it can feel like we are walking through a valley shrouded in despair. Yet, even in these moments of profound sadness, the Bible offers a beacon of hope, reminding us that grief, while a natural response to loss, can also be a pathway to deeper faith and renewal. 1 Thessalonians 4:13-14 does not diminish the reality of our pain but encourages us to grieve with the hope that comes from our faith in Christ. "Brothers and sisters, we do not want you to be uninformed about those who sleep in death so that you do not grieve like the rest of mankind, who have no hope." This passage reassures us that our grief is not a sign of weak faith but a natural human emotion that can coexist with a profound hope in God's promises.

Understanding grief through a biblical lens allows us to see it as a process—a journey through various stages where each phase contributes to healing and growth. The stages of grief, often described as denial, anger, bargaining, depression, and acceptance, are not linear but are aspects of a complex process that molds and shapes our spiritual and emotional lives. Faith infuses these stages with a unique perspective that transforms our inner turmoil into opportunities for deepening trust in God's sovereignty and compassion. For instance, in the anger or bargaining stages, you might question God's plan, a response that can feel unsettling for a believer. However, this is also a moment where faith can guide you to a deeper dialogue with God, where raw emotions can lead to profound truths about His nature and our reliance on Him.

When walking in faith, each stage brings its lessons and comforts. In the depths of depression that often accompany grief, the Psalms offer solace and understanding. Psalms like Psalm 34:18, "The Lord is close to the brokenhearted and saves those who are crushed in spirit," remind us that God's presence is near, especially in our lowest points. This scriptural truth can be a soothing balm, providing the strength to endure and gradually move toward acceptance, not just of the loss itself but of the new reality that emerges in its wake.

Moving forward after a loss is one of the most challenging aspects of the grieving process. Here, faith acts as both a shield and a compass—protecting against despair and guiding toward a future still in God's hands. Leaning on God's strength becomes essential during times when personal strength falters. Philippians 4:13 says, "I can do all this through him who gives me strength," it becomes not just a verse of empowerment but a daily lifeline. Moreover, faith can inspire us to channel our experiences of grief into helping others who may be navigating similar pains. Whether through sharing your own story of loss and recovery or simply being present for someone in their grief, these acts of compassion forge connections that can help heal broken hearts, including our own.

The stories of those who have traversed the terrain of loss with faith affirm the transformative power of trusting God's unwavering support. Consider the testimony of a widow who found renewed purpose in volunteering with grief support groups, using her sorrow journey to offer others empathy and hope. Or the story of a parent who, after the loss of a child, began a ministry to help other grieving parents find solace and meaning in their profound loss. These real-life testimonies speak of recovering from grief and growing through it, illustrating how faith can turn the most painful losses into pathways of new purpose and deepened faith.

In embracing grief with faith, we open ourselves to God's sustaining grace and the communal bonds that our shared experiences of loss can fortify. This approach does not promise an easy path; instead, it offers a meaningful one where each step, tear, and prayer leads us through grief and into a deeper embrace of life's fragility and God's eternal constancy. Grieving with hope is not about diminishing the pain of loss but about weaving this pain into the larger tapestry of our lives, marked by a faith that sustains, heals, and ultimately transforms.

OVERCOMING ANXIETY AND DEPRESSION THROUGH FAITH

In the quiet corners of our lives, where shadows loom and whispers of inadequacy reverberate, anxiety and depression often find their foothold. For many, these are not just passing clouds but persistent storms that darken their daily lives. Yet, within the pages of Scripture, we find not only acknowledgment of these struggles but also profound promises of peace and comfort. Verses like Psalm 34:17-18 reassure us, "The righteous cry out, and the Lord hears them; He delivers them from all their troubles. The Lord is close to the brokenhearted and saves those crushed in spirit." Here, the Bible doesn't shy away from acknowledging the depth of human despair. Instead, it assures us of God's attentive presence and unwavering support.

Philippians 4:6-7 further instructs us, "Do not be anxious about anything, but in every situation, by prayer and petition, with thanksgiving, present your requests to God. And the peace of God, which transcends all understanding, will guard your hearts and minds in Christ Jesus." This passage doesn't just offer a reprieve from anxiety; it provides a transformational strategy that involves prayer, gratitude, and the supernatural peace of God as a guard over our innermost thoughts and feelings.

Integrating Faith with Mental Health Practices

Recognizing that spiritual health is intricately connected to mental and emotional well-being, it becomes clear that integrating our faith with mental health practices is not just helpful; it's necessary. For Christians facing mental health challenges like anxiety and depression, incorporating faith into therapy can offer additional layers of understanding and avenues for healing. This holistic approach acknowledges the complete person—spirit, mind, and body—and leverages the unique strength of spiritual resources to complement psychological care.

For instance, Christian counseling often uses biblical principles alongside therapeutic techniques to address the root causes of anxiety and depression. Counselors may integrate prayer and Scripture reflection into sessions, providing cognitive behavioral strategies and spiritual practices that can reinforce healing. This integration helps individuals align their thought patterns with biblical truths, combating the lies and fears that often fuel anxiety and depression.

Practical Tools for Managing Anxiety and Depression

Practical tools and activities can serve as vital components of a comprehensive care strategy for managing anxiety and depression. Prayer meditation, for instance, involves focusing the mind and heart on Scripture or God's attributes, which can significantly reduce stress and promote a sense of peace. This practice can be enhanced by memorizing Scriptures that speak directly to feelings of anxiety and depression, such as Isaiah 41:10 or 1 Peter 5:7, embedding God's truth deeply within one's consciousness.

Creating a daily gratitude journal is another practical tool for shifting focus from worries and fears to appreciation and thankfulness. Writing down blessings daily can transform one's outlook, highlighting God's ongoing provision and care, which may otherwise go unnoticed in the turmoil of mental struggles.

Community and Isolation in Mental Health

The role of community in mental health cannot be overstated, especially given the isolation that often accompanies anxiety and depression. Engaging with supportive church groups or faith-based mental health resources can provide critical social support, reducing feelings of loneliness and despair. These communities offer safe spaces to share struggles and receive prayer, encouragement, and practical assistance. For example, many churches host support groups specifically for those dealing with mental health issues, offering both spiritual guidance and community connection.

Moreover, these groups often facilitate discussions around mental health from a biblical perspective, helping to destigmatize these issues within the church context. By fostering an environment where mental health is openly discussed, the church can play a pivotal role in breaking down the barriers of isolation and encouraging those suffering to seek help and community support.

In the ebb and flow of life's challenges, where anxiety and depression can sometimes take hold, the integration of faith and mental health practices provides a beacon of hope and a toolbox of resources. By rooting our strategies for overcoming these struggles in the wisdom and promises of Scripture, supported by practical tools and community engagement, we can travel through even the darkest moments, guided by the light of Christ's enduring presence and peace, according to His Word.

As we close this chapter on overcoming life's mental and emotional challenges through faith, we are assured that we are not alone in our struggles. The biblical insights, integrated practices, and community support discussed here are more than just concepts; they are part of a living faith that actively sustains us. As we turn our attention to the next chapter, we explore how our faith supports us internally and empowers us to engage with the world, transforming our trials into testimonies of hope and resilience.

CHAPTER ELEVEN

ENGAGING WITH THE WIDER WORLD THROUGH FAITH

Hebrews 11:1

"Faith is the substance of things hoped for, the evidence of things not seen."

Have you ever felt compelled to stand up for what is right yet wondered how your efforts could make a difference in this vast world? It's a common dilemma, especially when the issues seem overwhelmingly complex or entrenched. This is where faith-based advocacy emerges, bridging the gap between spiritual convictions and societal change. In this chapter, we'll explore how your faith can be a driving force in advocating for justice, mercy, and compassion, transforming your heart and the world around you.

During my quiet times and scriptural meditation, I often pray Hebrews 11:1 in my spirit by interchanging the word "faith" with "Jesus." Now, "Jesus" is the substance of everything we hoped for; He is the evidence of things I have never seen before. Faith is a supernatural dynamic that must land somewhere, and Christ Jesus is the target where we softly land.

FAITH-BASED ADVOCACY: INFLUENCING SOCIETY POSITIVELY

Understanding Faith-Based Advocacy

Faith-based advocacy is about putting our faith into action. It's about using our voices, resources, and time to influence policies and public opinion for the betterment of society. This form of advocacy goes beyond charity, addressing the root causes of issues and seeking to transform systems and structures in line with biblical mandates. One such mandate is found in Micah 6:8, which calls us to "act justly and to love mercy and to walk humbly with your God." This Scripture isn't just a call to personal holiness; it's a directive to engage actively with the world, promoting justice and mercy in our communities and beyond.

This kind of advocacy recognizes that faith is a powerful motivator for change. When you advocate from a faith-based perspective, you're driven by a vision of what the world could be—what it should be—according to divine standards of justice and love. It's about seeing the face of God in every person and acting against injustices that mar His creation. Whether fighting against poverty, standing up for the rights of the oppressed, or advocating for environmental stewardship, faith-based advocacy calls for a committed response to God's call for justice and peace.

Strategies for Effective Advocacy

Effective advocacy requires more than good intentions. It requires strategic actions, wise use of resources, and a clear voice. First, educate yourself thoroughly on the issues you care about. Understanding the difficulties of these issues allows you to speak knowledgeably and persuasively, whether you're discussing them in your community, with lawmakers, or through social media platforms.

Next, build alliances. Advocacy is more promising when collaborating with others who share your values and goals. Look for organizations, churches, or community groups already engaged in advocacy efforts and join forces to amplify your impact. You can coordinate actions, share resources, and support each other's campaigns.

Another critical strategy is to engage with policymakers. This can involve writing letters, making phone calls, or scheduling meetings with your representatives to discuss important issues. Be clear, concise, and respectful, but also persistent. Remember, policymakers are there to serve the public, and your advocacy ensures they are aware of and accountable for issues that matter to their constituents.

Case Studies of Successful Christian Advocacy

Consider the case of a Christian group that successfully lobbied for national policy changes to protect vulnerable children better. They brought significant legal reforms through public awareness campaigns, direct engagement with legislators, and collaboration with other advocacy groups. They succeeded in changing laws and raising societal awareness about these children's issues.

Another inspiring example is a church community that tackled homelessness in their city. They used their voices to advocate for more affordable housing and better support services. By partnering with local government and non-profits, they were instrumental in developing a new housing initiative that provided shelter and integrated support services to address the root causes of homelessness.

Balancing Advocacy with Humility and Respect

While it's essential to be passionate and persistent in advocacy, it's equally vital to approach these efforts with humility and respect. This means listening to others, even those with whom you disagree, and being willing to adjust your strategies based on new information or perspectives. It also involves respecting the dignity of all parties involved and avoiding tactics that dehumanize or demonize others.

True faith-based advocacy seeks transformation, not just of policies but of hearts—including ours. As you engage in advocacy, let it be a process that molds you into a more compassionate, just, and humble follower of Christ. This approach makes your advocacy efforts more genuine and likely to foster honest and lasting societal change.

In this exploration of faith-based advocacy, we see how deeply intertwined our spiritual lives are with our societal engagement. Your faith is not a passive state of being; it's a dynamic call to action, a mandate to bring about transformative change in the world. As you move forward, consider how you might use your voice and resources to advocate for justice and mercy, remembering that each effort, no matter how small, is a step toward a more just and compassionate world.

ENVIRONMENTAL STEWARDSHIP AS A FORM OF WORSHIP

When you step outside and breathe in the fresh air, feel the sun's warmth, or hear the rustle of leaves, are you reminded of the Creator's handiwork? Our calling to care for the Earth is not just a duty; it's an act of worship, an expression of our reverence and obedience to God. The Bible lays a foundational mandate for this stewardship in Genesis 1:28, where God commands humanity to "fill the earth and subdue it; have dominion over the fish of the sea, the birds of the air, and every living thing that moves on the earth." This scriptural charge is not about exploiting nature but managing it wisely, ensuring that God's creation is treated with respect and care. Understanding this role as stewards is essential, especially today, as environmental issues become more urgent and complex.

Embracing our role as caretakers involves more than just enjoying nature—it requires active engagement in preserving it. Simple actions like recycling, conserving water, or choosing sustainable products are ways you can practice environmental stewardship daily. Though these actions may seem small, they reflect a commitment to preserving God's creation for future generations. Moreover, by supporting larger-scale ecological initiatives—such as reforestation projects or clean water programs—you participate in environmental efforts to restore and maintain the health of the planet that God gave to us to steward.

The church has a significant role to play in this arena. Beyond individual actions, congregations can lead by example, implementing green practices in their operations and facilities. This could mean installing energy-efficient lighting, promoting recycling programs, or using eco-friendly materials during renovations. Churches can also host educational events to raise awareness about environmental issues, providing a platform for experts to discuss topics like climate change, biodiversity, and resource conservation. These initiatives make the church's operations more sustainable and send a powerful message to the community about the importance of environmental responsibility.

Moreover, environmental stewardship can be a potent signal to the broader community. When non-believers see Christians actively caring for the environment, it challenges common stereotypes of religious indifference to the planet's health. It shows that faith and care for the world are not mutually exclusive but are deeply interconnected. This witness can open doors for conversations about religion, as it demonstrates the practical implications of biblical teachings on creation care. It's an opportunity to discuss how Christian faith informs and inspires a commitment to sustainable living, offering a holistic approach to spirituality that values both the creator and His creation.

Engaging with environmental stewardship allows you to view each eco-friendly action as a practical duty and a meaningful expression of your faith. Whether you're planting trees, advocating for renewable energy, or teaching others about the importance of conservation, each effort is an act of worship, a testament to your commitment to obeying God's command to care for His creation. Through these actions, you affirm the value of every creature and ecosystem God has made, contributing to a legacy of stewardship that honors the Creator and sustains His creation for generations to come.

THE CHRISTIAN'S ROLE IN POLITICS AND PUBLIC POLICY

Stepping into politics as a Christian can often feel like navigating a complex labyrinth, where maintaining one's integrity and witness amid compromise and contention is a formidable challenge. Reflect on the dual call to be "in the world but not of the world." This biblical concept encapsulates the essence of Christian political engagement—it's about influencing society positively without losing our foundational values in the rough and tumble of political discourse. Engaging in politics as a Christian isn't just about casting votes or aligning with a party; it's active participation in shaping policies and public opinion to reflect the kingdom values of justice, mercy, and truth.

Navigating this engagement involves a delicate balance. On the one hand, you are called to be a voice for the voiceless and to defend the defenseless, advocating for policies that protect the vulnerable and promote justice. On the other hand, you must do so without compromising the principles that define your faith. This balance requires a deep understanding of political issues and an unwavering commitment to biblical principles. It also demands wisdom to know when to stand firm in your convictions and when to seek compromise for the greater good.

Now, let's delve into the principles guiding your political participation. At its core, Christian involvement in politics should be driven by the pursuit of the common good—this means seeking the welfare of the society in which you live, which Jeremiah 29:7 articulates as seeking the peace and prosperity of the city to which God has carried you. This pursuit is holistic, encompassing economic, social, and environmental well-being. It demands a commitment to promoting justice—rooted in the biblical understanding that God loves justice and commands us to do the same. But what does this look like in practice? It involves advocating for fair and equitable laws, supporting policies that uplift the marginalized, and opposing legislation that harms or exploits the least among us.

Furthermore, protecting the vulnerable is an unequivocal mandate for Christian political engagement. This principle is vividly illustrated in Proverbs 31:8-9, where we're urged to speak up for those who cannot speak for themselves, for the rights of all destitute. In political terms, this translates into championing the cause of groups such as people with low incomes, immigrants, the unborn, and older people—ensuring their rights are protected and their dignity upheld.

Consider historical and contemporary examples of Christians who have navigated the political arena without compromising their faith. Think of William Wilberforce, a devout Christian who fought tirelessly against the slave trade in Britain. His deep faith motivated his political actions, ultimately leading to the abolition of slavery in the British Empire. His life demonstrates how a steadfast commitment to Christ can bring profound societal changes. In more recent times, consider the numerous Christians involved in the civil rights movement in the United States, who were propelled by their faith to advocate for equality and justice, often at significant personal risk.

Engaging respectfully with opposing views is one of the most challenging aspects of political engagement. In a culture that often values confrontation over conversation, how can you, as a Christian, set a different tone? It begins with listening—truly listening—to understand rather than to refute. It involves engaging in discussions with grace and truth, standing firm on your convictions but doing so with a spirit of love and respect that reflects Christ. This approach not only fosters healthier and more productive political discussions but also serves as a witness to the transformative power of Christian engagement in public discourse.

In your journey through the political landscape, remember that your ultimate allegiance is not to a party or policy but to the Kingdom of God. This perspective will guide your decisions and actions and anchor you in the tumultuous seas of political engagement. As you advocate for laws and policies, do so as a citizen of heaven, bringing the values of justice, mercy, and humility to the public square. This is your calling and your challenge—as you engage, do so with the confidence that your efforts contribute to a larger narrative of redemption and restoration that God is unfolding in the world.

INTERFAITH DIALOGUE: OPPORTUNITIES AND CHALLENGES

In a world as richly diverse as ours, the ability to engage in meaningful dialogue with those of different faiths isn't just beneficial—it's essential. Interfaith dialogue involves open, respectful conversations between individuals of various religious backgrounds. Its fundamental purpose is to foster understanding and respect, recognizing that while we may hold divergent beliefs, we share a common humanity. Such interactions can demystify misconceptions, highlight shared values, and build bridges where walls might otherwise stand. The beauty of interfaith dialogue lies in its capacity to transform individual perspectives and forge pathways to greater societal harmony.

Preparing yourself for interfaith dialogue is akin to a significant exploration where every tool and insight counts. It begins with a deep knowledge of your faith in the scriptures. Knowing what you believe and why and standing on theological principles prepare you to share what you know clearly and confidently. However, this preparation isn't just about solidifying your views; it also involves approaching other faiths with respect and a gentle willingness to learn. This means studying other religions from credible sources and, whenever possible, directly from adherents of those faiths. It's about seeking to understand before seeking to be understood, approaching each conversation with the humility to know you do not need to have all the answers.

A reset of clear objectives is important. What do you hope to achieve through this dialogue? Is it to foster mutual respect, to collaborate on community projects, or to educate one another about your different beliefs? Clear objectives help steer conversations and ensure the dialogue remains constructive and focused on shared goals. These objectives also serve as a reminder of the larger purpose behind these conversations—building a community where diverse beliefs can coexist in harmony.

Navigating the challenges of interfaith dialogue requires tact and wisdom. Given the distinct nature of different religions, theological disagreements are inevitable. However, these discussions need not devolve into disputes. Framing these differences as opportunities for deeper understanding can transform potential conflicts into moments of insight. Another challenge is the risk of relativism, where the distinct truths of each faith might be downplayed to avoid conflict. Maintaining the integrity of your own beliefs while fully respecting others' convictions is a delicate balance but essential for meaningful dialogue.

Successful interfaith initiatives offer compelling models for implementing these principles. Consider a community where Christian and Muslim groups came together to address homelessness. By focusing on this common social issue, they not only improved conditions for many people but also deepened their understanding of each other's faith practices and humanitarian commitments. Another example is an interfaith youth camp that brings together teenagers from different religious backgrounds to participate in workshops and community service projects. These initiatives foster mutual respect among participants and build a foundation for lifelong cooperation and friendship.

In each case, the success of interfaith initiatives hinged on mutual respect, a focus on common goals, and a commitment to understanding each other's perspectives. These projects did not require participants to compromise their beliefs but to recognize the value of diversity and the strength that can be drawn from collective action. Interfaith dialogue and cooperation can lead to significant community betterment and contribute to a more inclusive and respectful society by focusing on what can be achieved together.

As we wrap up this exploration of interfaith dialogue, remember that the goal isn't to blur the lines that define our beliefs but to respect those lines while recognizing the vast common ground we share as members of the global community. This dialogue is an ongoing process, a continuous journey of learning, understanding, and cooperating that enriches our lives and the fabric of our communities. It's a testament to the power of faith not only to divide but, more profoundly, to unite. As you move forward, may your engagements be marked by deep respect, genuine curiosity, and a heartfelt commitment to fostering peace and understanding across the diverse tapestry of faith traditions.

CHAPTER TWELVE

THE FUTURE OF FAITH IN A RAPIDLY CHANGING WORLD

Psalm 121:1-2

"I lift my eyes to the mountains— where does my help come from? My help comes from the Lord, the Maker of heaven and earth."

Have you ever paused to consider how rapidly technology is evolving and shaping the very fabric of our society? From smartphones to smart homes, technological advances are not just conveniences; they are redefining how we live, interact, and even think. As Christians in this ever-evolving digital age, we face unique challenges that our forebears could scarcely have imagined. This chapter delves into these challenges, particularly the ethical dilemmas that arise with technological advancements such as artificial intelligence (AI), genetic engineering, and virtual reality. How we respond to these challenges reflects our faith and shapes our witness in a world increasingly reliant on technology.

PREPARING FOR FUTURE ETHICAL DILEMMAS IN TECHNOLOGY

Galatians 5:22-23

"But the fruit of the Spirit is love, joy, peace, forbearance, kindness, goodness, faithfulness, gentleness and self-control. Against such things, there is no law."

Identifying Potential Ethical Dilemmas

As technology advances at a breakneck pace, Christians are often left pondering the implications of these new tools and capabilities. Take artificial intelligence, for instance. AI can optimize efficiency and solve complex problems quickly, which is a boon for industries and services. However, it also raises significant ethical questions. What happens when AI begins to make decisions that humans traditionally made? How do we ensure these decisions are just and fair? Similarly, genetic engineering offers tremendous potential in treating and preventing diseases but brings up critical concerns about the sanctity of life and the natural order God has created.

Virtual reality, another frontier in technology, allows users to immerse themselves in digitally created worlds. This can be an incredible tool for education and understanding; imagine walking through the streets of ancient Jerusalem at the time of Jesus. Yet, it poses risks such as escapism or blurring lines between God's creation and artificial realities. Each of these technologies brings a set of complex ethical dilemmas that require us to tread with wisdom, discernment, and a deep anchoring in our biblical values.

Genesis 11:6

"The Lord said, 'If as one people speaking the same language, they have begun to do this, then nothing they plan to do will be impossible for them.'"

> *This verse, spoken in the context of the Tower of Babel, shows the potential of unified human endeavor. Modern technology is a testament to what we can achieve collectively. However, these efforts can lead to pride and self-sufficiency without God's guidance. It's a reminder that we should pursue technological advancements while remaining humble and reliant on God.*

— *PASTOR EMILY DAVID*

Biblical Principles for Technological Ethics

In navigating these dilemmas, we are not left to our own devices. Scripture provides foundational principles that can guide our engagement with technology. The sanctity of life, a principle woven throughout Scripture from the creation narrative through to the teachings of Jesus, reminds us to uphold the value and dignity of all life in our engagement with technologies like AI and genetic engineering. The importance of truth and transparency, emphasized in Proverbs 12:22 and Ephesians 4:25, guides us in advocating for technologies that enhance truthful communication and discourage deceit. Lastly, the stewardship of creation entrusted to humanity in Genesis calls us to use technology in ways that care for and preserve the world around us, not exploit it.

Engaging With Tech Communities

Christians must understand and reflect on these ethical challenges and engage actively in the spaces where technology decisions are made. This means advocating for the presence and voice of Christian ethicists and technologists in tech companies, research institutions, and policy-making arenas. By participating in these discussions, Christians can influence the development and implementation of technologies that align with biblical values, promoting ethical practices that respect human dignity and God's creation.

Case Studies of Ethical Decision-Making

Consider the story of a Christian software developer involved in a project to develop an AI-driven hiring tool. Aware of the potential for AI to reflect and even amplify human biases, he advocated for the incorporation of mechanisms to ensure fairness and transparency in the AI algorithms. His engagement ensured the tool was used to enhance, not hinder, fair employment practices.

Another case involves a Christian geneticist who steered the complex ethical terrain of gene editing. Faced with the potential to correct genetic defects before birth, she wrestled with the implications of "playing God" with human genetics. Her approach focused on treatments for debilitating diseases rather than enhancements or alterations, aiming to alleviate suffering while respecting the integrity of the human genome as God's creation.

These examples show how Christians can navigate the ethical challenges posed by modern technology by rooting their behavior in biblical principles and actively participating in the broader tech community. By doing so, believers contribute to a future where technology is used not just for innovation and profit but also for the benefit of others and to the glory of God.

THE NEXT GENERATION: INSTILLING FAITH IN YOUTH

Have you noticed how the vibrant energy of youth can invigorate a community? Yet, maintaining this vibrancy and ensuring it embraces profound spiritual truths is one of the significant challenges facing today's church. The younger generations, often called Gen Z and Gen Alpha, are growing up in a world vastly different from twenty years ago. They traverse landscapes filled with secularism, where relative truths override absolute standards, and digital platforms bombard them with much information and many viewpoints. This environment makes retaining faith a complex journey for many young people, who often question the relevance of church and Christianity in their lives.

Addressing these challenges requires a dynamic shift in how youth ministry operates. It's no longer just about providing a weekly youth group meeting. Today, the ministry must engage with young people on platforms and in languages that resonate deeply with them. This might mean leveraging social media as an announcement tool and a dialogue and spiritual engagement platform. Picture a youth group that conducts Bible study discussions over Instagram Live or shares testimonial snippets via TikTok—these are ways the message of Christ can transcend traditional boundaries and tap directly into the digital spaces where young people spend much of their time.

Moreover, integrating technology means more than just being present on social media. It means using these tools to create interactive and engaging content that challenges the youth to think and reflect on their faith. For instance, virtual reality experiences that allow young people to walk through biblical landscapes or augmented reality apps that provide in-depth explanations of biblical sites or events can transform how they perceive and understand Scripture. By bringing the Bible to life this way, youth ministry can provide a fresh perspective that enhances engagement and retention of scriptural truths.

However, technology alone isn't the answer. Authentic relationships and mentorship are vital in guiding young people. They often look for role models who preach faith and live it out authentically. This is where mentorship plays a pivotal role. The church can foster meaningful relationships that encourage spiritual maturity by pairing young individuals with older, more experienced Christians who can share life experiences, wisdom, and spiritual insights. These mentors are tangible proof that faith can thrive amid modern challenges, providing a relatable blueprint for navigating life's problems with God at the center.

Success stories from various ministries around the globe serve as a testament to the fruitfulness of these approaches. Take, for example, a youth ministry in urban Chicago that restructured its approach by integrating hip-hop and spoken word into its programs. They provided a platform for young people to express their faith creatively while addressing social issues pertinent to their context. This approach increased engagement and helped the youth understand how their faith intersects with cultural expression and social justice.

Another success story comes from a small community church in rural Texas, which introduced a mentorship program linking its youth with Christian business leaders in the community. This program guided the youth in their faith journeys. It equipped them with practical skills for their future careers, illustrating the multifaceted role of mentorship in developing spiritual and professional capacities.

In each of these examples, the key to success was the ministry's ability to adapt and resonate with the interests and needs of the younger generation, all while grounding their efforts in biblical truth and authentic relationships. As we continue to explore the dynamics of faith engagement among youth, it becomes increasingly clear that our approaches must be as adaptable and vibrant as the young people we aim to reach. Through innovative uses of technology, culturally relevant teaching methods, and the irreplaceable value of human mentorship, we can offer a compelling vision of what it means to follow Christ in today's world. This vision is heard, seen, and experienced in ways that stir the heart and spirit of the next generation.

THE GLOBAL CHURCH: EXPANDING BEYOND BORDERS

As you sit in your local church or perhaps scroll through a digital sermon on your tablet, it's exhilarating to ponder that Christianity, a faith rooted in the teachings of a first-century carpenter from the small town of Nazareth, now echoes in every corner of the globe. Recent trends indicate a significant shift in the demographic heart of Christianity to the Global South—Africa, Latin America, and parts of Asia. This movement isn't just a numerical shift; it represents a dynamic transformation in the cultural expressions and theological emphases of global Christianity. For centuries, the West was the epicenter of Christian thought and practice. However, as the gospel takes deeper root in the soils of the Global South, it's blossoming into expressions that are both refreshing and challenging to traditional Western paradigms of worship and ministry.

This geographical and cultural shift invites us to reevaluate and reimagine the scope and methods of our mission work. Historically, Western missionary efforts, though well-intentioned, often carried with them a shadow of cultural imperialism, where Western customs, practices, and interpretations of Christianity were imposed on other cultures. Today, there's a growing recognition of the richness and validity of Indigenous expressions of the Christian faith, which respect and incorporate local customs and wisdom. This respects the diversity of the body of Christ and enriches the entire church. Embracing a partnership model in missions—in which Western Christians support rather than lead, where they participate as learners and teachers—is critical. Such an approach not only prevents the pitfalls of paternalism but also fosters mutual respect and enriches all involved through the exchange of spiritual gifts and insights.

Moreover, the digital age offers unprecedented opportunities for global church members to connect, collaborate, and learn from one another. Digital platforms break down geographical barriers, allowing a pastor in Nairobi to share insights with a congregation in Nicaragua or a Christian in rural India to participate in a Bible study hosted from Canada. The potential for these connections to foster a truly global and interconnected church is immense. Through online conferences, worship services, and training programs, believers worldwide can access resources, training, and encouragement that would have been out of reach just a few decades ago. This digital connectivity facilitates education and spiritual growth and allows Christians living in restricted or hostile environments to receive support and fellowship from the global church body.

Case Studies of International Church Partnerships

Reflect on the story of a partnership between a church in Norway and a community in Rwanda. The Norwegian church initially engaged with the intent to help, sending funds and resources to support local projects. However, as the relationship developed, it transformed into a genuine partnership. The Rwandan church invited their Norwegian brothers and sisters to partake in local leadership training programs deeply rooted in the community's cultural context and daily realities. This exchange was a two-way street, where the Norwegians found their understanding of discipleship and community engagement profoundly deepened and enriched by the Rwandan perspective.

Similarly, consider a collaborative project between churches in the Philippines and Australia. They joined forces to address environmental degradation, a pressing issue affecting communities in both regions. By combining resources, human capital, and local knowledge, they developed community programs that tackled environmental challenges and empowered local churches to take leadership roles in ecological stewardship. This partnership highlighted how shared global challenges could become platforms for cross-cultural collaboration and mutual growth.

These stories illuminate the beautiful potential of cross-cultural partnerships that honor local cultures and foster reciprocal relationships. As we continue to acknowledge the influences of a global church, let us strive for a model of mission and ministry that embodies respect for others, partnership, and the building up of other believers. The future of Christianity is global, and in this tapestry of diverse cultural expression, every thread is essential, every hue vibrant, and every pattern rich with the potential to teach us more about the manifold wisdom of God. In this global fellowship, as we share our burdens, joys, and understanding of the gospel, we move closer to the scriptural vision of every tribe, tongue, and nation united in worship before the throne of God.

CONTINUING PERSONAL FAITH JOURNEYS IN AN UNCERTAIN WORLD

In a world where the only constant is change, how can you, as a believer, cultivate a faith that not only endures but thrives? Amid rapid societal shifts, economic fluctuations, and political upheavals, building resilience in your faith is more important than ever. It involves anchoring yourself in truths that transcend the shifting sands of culture and circumstance. Resilience in faith doesn't mean stubbornness in old ways; instead, it's about developing flexibility that allows you to respond to change with spiritual maturity and wisdom.

One of the key ways to build this resilience is by embracing continuous learning and adaptation in your spiritual journey. This means not just sticking to what you already know or feel comfortable with but actively seeking to understand how God's word applies to the complexities of modern life. It involves engaging with Scripture profoundly and regularly and reaching beyond to understand theological perspectives and interpretations that challenge and expand your own. For instance, studying how the early church waded through cultural shifts offers insights into handling today's changes. Engaging with contemporary issues through a biblical lens - whether discussions about social justice, environmental stewardship, or technological ethics—helps you apply your faith in practical, relevant ways.

Adapting traditional spiritual disciplines to modern contexts is another vital aspect of this journey. Consider, for example, the practice of solitude and silence, which are counter-cultural disciplines in our noisy, always-connected world. Finding time for silence in your daily routine by starting each day with a few moments of quiet reflection instead of immediately checking your phone can help cultivate a sense of peace and presence that strengthens your faith. Similarly, fasting, traditionally understood as abstaining from food, can be expanded to digital fasting—taking breaks from social media and digital consumption to focus on spiritual nourishment.

The power of personal testimonies in this context cannot be understated. Hearing how others choose to believe in God through crises can be incredibly encouraging. Imagine the story of a young entrepreneur who faced business failure during an economic downturn. Instead of succumbing to despair, his faith led him to view this setback as a redirection, eventually guiding him to a path of impactful community service. Or consider a family that maintained a strong faith while contemplating the possibilities of political sanctuary. Their story of trust in a sovereign God and perseverance in the face of instability can offer powerful inspiration and practical lessons for maintaining faith amid life's storms.

These stories encourage us and illustrate how resilience, continuous learning, and adapted spiritual disciplines play out in everyday life. They remind us that we are not alone in our struggles and that our faith journey is a shared experience enriched by the stories of those who walk alongside us.

In this ever-changing world, your faith journey is personal and communal, a path of individual growth and collective journeying. As you navigate the uncertainties of life, remember that building resilience in your faith, embracing continuous learning, adapting spiritual disciplines, and drawing strength from personal testimonies are not just strategies but invitations. They invite you to deepen your relationship with God, to engage with the world thoughtfully and confidently, and to contribute to the faith journeys of others just as they contribute to yours. This dynamic interplay of learning, adapting, and overcoming ensures that your faith is not just a static relic but a living, breathing aspect of your daily life.

LET FAITH WIN OVER FEAR

As you close the final pages of this book, I hope your faith in God is more robust, deeper, and a more significant part of your daily life. In this book, I aim to help you strike the right balance between technology, worship, and forming part of a faith-based community. Thankfully, the rapid advancement of technology isn't something to fear. Instead, many faith-centered tools can make incorporating prayer, meditation, and faith-based contemplation into our daily lives easier. If this book has helped you adapt your faith-based practices to these technologically savvy times, I hope you can share your opinion.

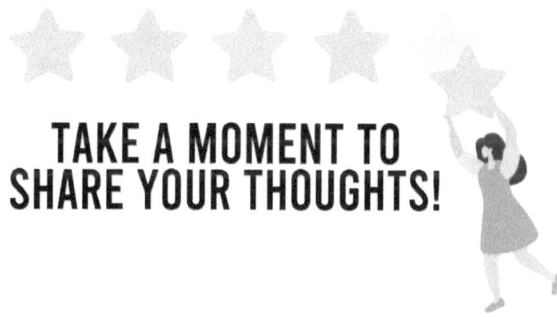

TAKE A MOMENT TO
SHARE YOUR THOUGHTS!

Thanks for your support. The word of God was always meant to be shared so that more people can know that God walks by their side.

Scan the QR code below.

SCAN ME

CONCLUSION

Proverbs 16:3

"Commit to the Lord whatever you do; he will establish your plans."

As we wrap up our journey through *The Divine Initiative & the Human Response: Navigating Your Faith in the Digital Age*, I hope you've found comfort and challenge in the shared pages. We've explored the majestic interplay of divine sovereignty and human response, especially as these divine moments intersect with our digital era's rapid advancements and potential pitfalls.

Scripture has been our unfailing roadmap throughout this exploration. It has offered us deep insights and practical applications that are as relevant today as they were in the Apostles' time. The Bible remains our essential guide in understanding how to live faithfully in God's kingdom, whether through the complexity of Joseph's trials or the daily digital interactions we wrestle with.

The stories, personal anecdotes, and testimonials woven into each chapter have highlighted the rich tapestry of faith experiences across different lives and circumstances. Each story uniquely expresses a universal truth—we are all navigating this journey of faith, each with our unique battles, breakthroughs, and moments of beauty.

Every chapter has incorporated interactive elements—reflection questions, prayer prompts, and action steps. I hope these have been exercises and gateways to deeper engagement with your faith in God and your surrounding communities.

We've tackled some of the pressing challenges of our time—navigating faith amid the digital deluge, discerning the noise of social media, and finding genuine fellowship online. These discussions were not just theoretical; they were anchored in fundamental strategies to help you live out your faith in the modern world.

Our discussions on divine sovereignty and human free will aimed to demystify these profound theological concepts, bringing them down to earth through relatable stories and guided devotionals. Seeing God's hand in the daily grind is possible, and I hope this book has illuminated that path for you.

As you move forward, I urge you to apply the actionable strategies we've discussed in your life. Whether setting boundaries for digital consumption, engaging in local community service, or deepening your scriptural meditation, each step moves toward living a vibrant and actionable faith.

The Christian community is both a gift and a necessity for spiritual growth. I encourage you to seek out or invest more in these divine connections. They are your support network, spiritual family, and co-travelers on this faith journey.

Life's challenges can be daunting, but remember, you are equipped. The biblical strategies outlined in this book are your armor and tools. Engage deeply with Scripture, lean into your community, and let your life be a testimony of faith in action.

Now, I call you to action. Take what you have learned about divine sovereignty and human choice, about living faithfully in a digital age, and put these insights into practice. Be the light in your online and offline worlds, showing the love and wisdom of Christ in all you do.

Finally, my hope and prayer for you is that you grow ever more in understanding who you are in Christ. May you live each day with the grace and wisdom from above, using every challenge to deepen your faith in the one who authored it (Hebrews 12:1-2). May your interactions be fruitful, your community ties be strong, and your life reflect God's divine love, which calls us all to respond to His goodness and grace.

Let's continue this journey, exploring, growing, and living out the profound truth that we are called to respond while God indeed initiates. God bless you richly as you continue to walk in His way, truth, and life.

REFERENCES

Six ways to effectively build community in your church plant. (n.d.). Converge. https://www.converge.org/article/6-ways-to-effectively-build-community-in-your-church-plant

10 great tools for your spiritual growth | af news. (n.d.). Amazing Facts. https://www.amazingfacts.org/news-and-features/news/item/id/10930/t/10-great-tools-for-your-spiritual-growth

20 calming bible verses about patience to help you get through the day. (2020, March 30). Country Living. https://www.countryliving.com/life/g31967106/bible-verses-about-patience/

30 Important Bible Verses About Technology (2024, July 16). *Pastor Emily David*https://explainingthebible.com/bible-verses-about-technology/

50+ bible verses for healing—Powerful scripture quotes. (n.d.). Bible Study Tools. https://www.biblestudytools.com/topical-verses/healing-bible-verses/

Abraham's bind: & other bible tales of trickery, folly, mercy and love. (n.d.). Turner Publishing. https://turnerpublishing.com/products/abrahams-bind

Ai and christianity: Navigating the intersection of technology and faith in ministry work. (n.d.). https://www.delmethod.com/blog/ai-and-christianity

Altruism. (n.d.). Effective Altruism for Christians. https://www.eaforchristians.org/the-call-to-altruism

Annan, K., & Aten, J. (n.d.). *The challenges and benefits of moving church online.* The Better Samaritan with Jamie Aten and Kent Annan. https://www.christianityto day.com/better-samaritan/2020/july/challenges-and-benefits-of-moving-church-online.html

Biblical financial principles and 5 practical steps to live by them. (2023, June 22). Christian Stewardship Network. https://www.christianstewardshipnetwork.com/blog/2023/6/22/biblical-financial-principles-and-5-practical-steps-to-live-by-them

Bloom, J. (2018, April 6). Unanswered prayers are invitations from god. *Desiring God.* https://www.desiringgod.org/articles/unanswered-prayers-are-invitations-from-god

Christian Reformed Church (n.d.). *Biblical advocacy 101.* https://www.crcna.org/sites/default/files/36318_osj_advocacy_brochure_us_web.pdf

Churchtrac blog. (n.d.). ChurchTrac. https://www.churchtrac.com/articles

Dunker, M. P. (2023, December 19). 8 keys to a more powerful prayer life. *World*

Vision. https://www.worldvision.org/christian-faith-news-stories/keys-power ful-prayer-life

Ephesians 5:21-33 bible study—Willing submission & the bride of christ. (n.d.). https://jesusplusnothing.com/series/post/ephesians-5-bible-study-willing-submission-bride-christ

Evans, G. A. S. and J. (2024, March 15). Christianity's place in politics, and 'Christian nationalism.' *Pew Research Center.* https://www.pewresearch.org/religion/2024/03/15/christianitys-place-in-politics-and-christian-nationalism/

Faith in american public life. (n.d.). *Baylor University Press.* https://www.baylorpress.com/9781481309707/faith-in-american-public-life

God's sovereignty and human free will. (n.d.). Focus on the Family. https://www.focusonthefamily.com/family-qa/gods-sovereignty-and-human-free-will/

Graham, J. M. (2018). Edward t. Oakes, a theology of grace in six controversies. Grand rapids: Eerdmans, 2016, *International Journal of Systematic Theology, 20*(4), 579–583. https://doi.org/10.1111/ijst.12305

Harber, I. (2023, August 11). *Social media is a spiritual distortion zone.* The Gospel Coalition. https://www.thegospelcoalition.org/article/social-media-spiritual-distortion/

Hart, A. (2023, April 21). The top 5 best prayer apps for christians. *Global Christian Relief.* https://globalchristianrelief.org/christian-persecution/stories/best-prayer-apps/

HC. (2023, January 17). *Ethics in the age of AI: Defining and pursuing the good for our good and the good of our communities | Houston Christian university.* https://hc.edu/center-for-christianity-in-business/2023/01/17/ethics-in-the-age-of-ai/

How should Christians handle disputes (Matthew 18:15-17)? (n.d.). GotQuestions.Org. https://www.gotquestions.org/Christian-disputes.html

How to grieve as a christian. (2019, December 20). Gentle Reformation. https://gentlereformation.com/2019/12/20/how-to-grieve-as-a-christian/

Hutchings, T. (2011, January). *Online Christian churches: three case studies.* Journal for the Academic Study of Religion 23(3). https://www.researchgate.net/publication/270416668_Online_Christian_Churches_Three_Case_Studies

In all things charity: A pastoral challenge for the new millennium | usccb. (n.d.). https://www.usccb.org/resources/all-things-charity-pastoral-challenge-new-millennium

Integrating technology in ministry: Adapting for the digital age | plnu. (2024, March 18). https://www.pointloma.edu/resources/theology-christian-ministry/technology-ministry-adapting-digital-age

John. (n.d.). The holy spirit: Our compass in decision-making. *Flourish.* from https://reconciledworld.org/flourish/the-holy-spirit-our-compass-in-decision-making/

Johnson, C.M. (n.d.) *4D's of spiritual growth: the local church's role in spiritual growth and discipleship*. Asbury Theological Seminary. https://place.asburyseminary. edu/cgi/viewcontent.cgi?article=2467&context=ecommonsatsdissertations

Kshirsagar, N. (2023, October 19). Digital devotion: How technology enhanced spiritual practices. *Medium*. https://medium.com/@nikhil2050/digital-devo tion-how-technology-enhanced-spiritual-practices-7453edd5403b

Leung, J., & Li, K.-K. (2023). Faith-based spiritual intervention for persons with depression: Preliminary evidence from a pilot study. *Healthcare, 11*(15). https:// doi.org/10.3390/healthcare11152134

Life, A. (2022, July 8). 7 ways to overcome spiritual dryness. *Abundant Life*. https:// livingproof.co/7-ways-to-overcome-spiritual-dryness/

Marshall, S. (2018, November 1). 24 divine intervention examples that shows god's hand at work. *GodUpdates*. https://www.godupdates.com/24-divine-interven tion-examples-that-shows-gods-hand-at-work/

Member, A. S. (2024, June 25). *A biblical perspective on environmental stewardship*. Acton Institute. https://www.acton.org/public-policy/environmental-steward ship/theology-e/biblical-perspective-environmental-stewardship

Ortlund, G. (2017, June 16). *7 ways to fight distraction in prayer*. The Gospel Coalition. https://www.thegospelcoalition.org/article/7-ways-to-fight-distraction-in-prayer/

Piper, J. (2007, June 20). More thoughts for fathers on ephesians 6:4. *Desiring God*. https://www.desiringgod.org/articles/more-thoughts-for-fathers-on-ephesians-6-4

Radical obedience in ordinary life. (n.d.). The Daily Grace Co. https://thedailygraceco. com/blogs/the-daily-grace-blog/radical-obedience-in-ordinary-life-2

Scott, E. R. (n.d.). *The sustaining power of faith in times of uncertainty and testing*. Retrieved June 25, 2024, from https://www.churchofjesuschrist.org/study/eng/ general-conference/2003/04/the-sustaining-power-of-faith-in-times-of-uncer tainty-and-testing

See what takes place: Virtual reality in teaching about religious rituals. (2022, November 4). *Religion Matters*. https://religionmatters.org/2022/11/04/see-what-takes-place-virtual-reality-in-teaching-about-religious-rituals/

Taketa, C. (n.d.). *Why small groups?* Small Groups. https://www.smallgroups.com/ articles/2012/why-small-groups.html

Taylor, J. (2015, March 13). *22 benefits of meditating on scripture*. The Gospel Coalition. https://www.thegospelcoalition.org/blogs/justin-taylor/22-benefits-of-meditating-on-scripture/

The 20 best spiritual podcasts for growth and awakening. (2023, October 26). Podcastle Blog. https://podcastle.ai/blog/best-spiritual-podcasts/

The 'mad virtues' of a secular society. (2016, August 9). Catholic Exchange. https://catholicexchange.com/mad-virtues-secular-society/

The role of community in discipleship | cru. (n.d.). Cru.Org. https://www.cru.org/us/en/train-and-grow/help-others-grow/discipleship/the-role-of-community-in-discipleship.html

The sovereignty of God: Case studies in the Old Testament | monergism. (n.d.). https://www.monergism.com/sovereignty-god-case-studies-old-testament

Three bible heroes who doubted. (2012, August 9). *Bible Gateway Blog.* https://www.biblegateway.com/blog/2012/08/three-bible-heroes-who-doubted/

United States Institute of Peace. (n.d.). What works? Evaluating interfaith dialogue programs. https://www.usip.org/sites/default/files/sr123.pdf

Using tech well takes God-given discernment. (n.d.). Resurrection Church. https://resurrection.church/gps-guide/using-tech-well-takes-god-given-discernment/

What does Proverbs 13:20 mean? (n.d.). BibleRef.Com. https://www.bibleref.com/Proverbs/13/Proverbs-13-20.html

Willmington, H. (2017). *Jehovah-Jireh.* https://core.ac.uk/download/129594069.pdf

ALSO BY DWAINE AJ WHOGOES

THE DIVINE INITIATIVE AND THE HUMAN RESPONSE:

Navigating Your Faith in the Digital Age

STRATEGIES FOR HIGHER THINKING:

Enhancing Our Lives by Embracing AI

THE HUMAN-CANINE CONNECTION:

Enriching Your Life by Loving Your Dog